T0312925

---

# Insights on Education
# Reform in China

# RIVER PUBLISHERS SERIES IN INNOVATION AND CHANGE IN EDUCATION - CROSS-CULTURAL PERSPECTIVE

Indexing: All books published in this series are submitted to the Web of Science Book Citation Index (BkCI), to CrossRef and to Google Scholar.

Nowadays, educational institutions are being challenged as professional competences and expertise become progressively more complex. This is mainly because problems are more technology-bounded, unstable and ill-defined with the involvement of various integrated issues. Solving these problems requires interdisciplinary knowledge, collaboration skills, and innovative thinking, among other competences. In order to facilitate students with the competences expected in their future professions, educational institutions worldwide are implementing innovations and changes in many respects.

This book series includes a list of research projects that document innovation and change in education. The topics range from organizational change, curriculum design and innovation, and pedagogy development to the role of teaching staff in the change process, students' performance in the areas of not only academic scores, but also learning processes and skills development such as problem solving creativity, communication, and quality issues, among others. An inter- or cross-cultural perspective is studied in this book series that includes three layers. First, research contexts in these books include different countries/regions with various educational traditions, systems and societal backgrounds in a global context. Second, the impact of professional and institutional cultures such as language, engineering, medicine and health, and teachers' education are also taken into consideration in these research projects. The third layer incorporates individual beliefs, perceptions, identity development and skills development in the learning processes, and inter-personal interaction and communication within the cultural contexts in the first two layers.

We strongly encourage you as an expert within this field to contribute with your research and help create an international awareness of this scientific subject.

For a list of other books in this series, visit www.riverpublishers.com

# Insights on Education Reform in China

### Editors

## Zhiying Nian
## Qinhua Zheng
## Li Chen

Beijing Institute for Learning Society
Beijing Normal University, China

LONDON AND NEW YORK

**Published 2018 by River Publishers**
River Publishers
Alsbjergvej 10, 9260 Gistrup, Denmark
www.riverpublishers.com

**Distributed exclusively by Routledge**
4 Park Square, Milton Park, Abingdon, Oxon OX14
4RN 605 Third Avenue, New York, NY 10158

First published in paperback 2024

*Insights on Education: Reform in China* / by Zhiying Nian, Qinhua Zheng, Li Chen.

*Routledge is an imprint of the Taylor & Francis Group, an informa business*

Publisher's Note
The publisher has gone to great lengths to ensure the quality of this reprint but points out that some imperfections in the original copies may be apparent.

While every effort is made to provide dependable information, the publisher, authors, and editors cannot be held responsible for any errors or omissions.

ISBN: 978-87-93379-64-0 (hbk)
ISBN: 978-87-7004-421-9 (pbk)
ISBN: 978-1-003-33858-1 (ebk)

DOI: 10.1201/9781003338581

# Contents

# List of Contributors

**Dayong Yuan,** *Institute of Vocational and Adult Education (IVAE) at Beijing Academy of Educational Sciences (BAES), Beijing, China*

**Mingli Fan,** *Hebei University, Baoding, China*

**Qinhua Zheng,** *Beijing Normal University, Beijing, China*

**Xiaojie Zeng,** *Beijing Institute for the Learning Society, Beijing Normal University, Beijing, China*

**Yijin Zhang,** *Faculty of Education, Beijing Normal University, Beijing, China*

**Yimin Gao,** *Institute of International and Comparative Education, Beijing Normal University, Beijing, China*

**Yu Hong,** *Hebei University, Beijing, China*

**Zhiying Nian,** *Beijing Institute for the Learning Society, Beijing Normal University, Beijing, China*

# List of Figures

# List of Tables

# 1

## Introduction

**Yimin Gao**

Institute of International and Comparative Education,
Beijing Normal University, Beijing, China

In recent years, as indicated by the fact that Shanghai earned the top-ranking PISA results in the world, China's education system has attracted more and more attention. A glimpse at the scale of education in China – with more than 200 million students in school – reveals that this is an unusual case. But the main reason that China's education system has attracted the interest of researchers is that, since the late 1970s, China has initiated landmark reforms in education. The sheer scale of the education system today was, to a certain extent, brought about by these reforms.

During the 1980s, a tide of educational reform swept the globe. In China, however, this period of educational reform was driven primarily by internal politics. In 1976, China saw the end of 10-year "Cultural Revolution," an ironic upheaval that enforced an empty proposition of culture, in the sense that the government not only banned traditional social culture, but also rejected modern alternatives. As a consequence, the educational system, which had been informed by time-tested international models, was compressed in an unprincipled way: educational content largely gave way to national ideology, and the so-called "social practice," which is ordinarily separated from modern knowledge, became an important approach and means of education, resulting in a serious shortage of ordinary workers with modern ideology, creative talent and knowledge of modern science and technology. In response to this crisis, the CPC Central Committee issued *Decisions on the Reform of the Educational System* in 1985, negating the educational policies implemented during the "Cultural Revolution", and restoring the previous educational order with a view toward establishing an educational system that would adapt to modernization. This educational reform possessed its own logic and largely reflected the commonalities of educational development in different countries. As a result, China's education has becoming increasingly "internationalized" during the course of educational reform, meaning that China also needs to address educational problems that have been solved or are being solved in other countries. For example, China must achieve equal access to education gradually, which requires continuously expanding the scale of education to make it accessible to more people; enhance the quality of education, so that it satisfies the need of personal development and the demands of society and reflects new achievements of science and technology and human civilization; and make Chinese school systems more open than they have been historically, so as to accelerate the development of learning in society step by step. During this process, China has always actively paid attention to the experiences of other countries and has used them as points of reference for educational reform.

Recent years have seen an increase in the number of English-language books on educational reform in China. Unlike these books, which examine macroscopic policies, this book examines pre-school education, elementary education, vocational education, higher education, adult education, life-long learning, and other aspects of education, and only analyzes one specific problem in each of these contexts for the purpose of comparing China's educational reforms to their overseas counterparts through microscopic study. The authors are all young and middle-aged scholars who have made a mark in their own fields of study. Most of them have been awarded master's degrees or doctorates from Beijing Normal University, one of China's first institutions of higher education to train teachers, which boasts a distinguished tradition in education and research and enjoys a high reputation in academic circles.

In Chapter 1, Dr. Mingli Fan analyzes pre-school educational reforms in Hebei Province: marketization, for instance, is a keyword in China's economic reforms that has entered elementary education very rapidly, against all expectations. One outcome of marketization is that large numbers of quasi-public sectors, such as state-owned enterprises and public institutions with backgrounds in public finance, do not run kindergartens or nurseries any longer: the government has instead relaxed or even partially given up the management of pre-school education. This kind of marketization resulted in substantial damage to the public interest, particularly in the context of China's family planning policy. It is not difficult to imagine people's dissatisfaction with "high admission fees and a difficult admission experience" in pre-school education. Hebei Province has solved this problem relatively well and popularized three-year pre-school education in two-thirds of rural areas by taking a more prudent approach to marketization.

In Chapter 2, Xiaojie Zeng analyzes the problems associated with selecting a school in China, also in the context of marketization. Of course, under a planned economic system, the phenomenon of "selecting a school with honors or power and influence" is well established and regarded as a reasonable way of selecting excellent students to efficiently "cultivate talent quickly" and "cultivate outstanding talents." Children of leaders and cadres are able to study in high-quality schools as a reward for their contribution to the revolution, a policy that has never been questioned. However, the "fairness-oriented" educational reform of the new era has called into question the process of "selecting a school with honors or power and influence." The practice has not been entirely discontinued but has continued to evolve into a benefit exchange. The severe inadequacy of public input also provides certain "reasonableness" for "selecting a school with wealth." Ironically, economic marketization has actually boosted the rights-awareness of Chinese citizens. Persistent inequality in school selection has developed into a major social problem.

In Chapter 3, Yijin Zhang analyzes some cases of deregulation in local administrative educational departments, which is also considered a part of marketization. That the government is too powerful is one of the major problems to be solved through China's education reform. However, the deregulation of education, especially elementary education, is especially difficult. China's educational administrative system features a centralization of power that allows it to operate like an independent kingdom. However, against a backdrop of rapid expansion in the scale of education, gravely insufficient public resources, and more participating social forces, deregulation has become an inevitable trend. Previous modes of bureaucratic administrative management have given way to new "governing" modes such as government procurement of services. In this way, educational administrative power is being divided.

In Chapter 4, Dr. Dayong Yuan analyzes problems related to the poor recruitment performance of vocational education institutions in Beijing. Although "secondary technical schools" and "secondary specialized schools" belonging to the category of vocational education have existed for a long time, the concept of vocational education first appeared after the reform and opening-up policy. While there has been opposition to the marketization of elementary education, people usually agree that vocational education does need marketization, especially in order to be closely integrated with the labor market in specific industries. Yet this has proven very difficult to achieve.

In traditional Chinese cultural psychology, vocational education is mistrusted and thought of as inadequate, and consequently is not attractive to the market. The incomplete development of the labor market as a whole is one of the reasons for the underdevelopment of vocational education.

In Chapter 5, Dr. Yu Hong discusses the internal and external governance of contemporary Chinese universities. Some Chinese scholars find the formulation of the statutes of public universities ironic. Why? Because statutes must be conceived by organizers, and the sole organizer of Chinese universities is the Chinese government. In addition, universities already have specific management and operation regulations, so it is not necessary to set out general statutes. Moreover, universities actually have only limited decision-making powers – for instance, university statutes need to be approved by the Ministry of Education. But in any case, the setting of university statutes is an illustrative case through which to examine the management status of Chinese universities. The establishment of countervailing power structures such as Academic Committees, Teaching Committees, and Professor Committees reflects the sense in which China is still exploring the modern university system. The fact that the Chinese government insists on letting universities formulate these statutes can also be considered an effort to respect the decision-making power of universities and consolidate their self-government.

In Chapter 6, Dr. Qinhua Zheng describes the development of the Open University of China (OUC). OUC is a new venture for China. Its name alone suggests that it is inspired by international trends. However, in recent years, the situation of adult education in China has changed. As the scale of higher education rapidly expands, there are more opportunities for high school graduates to enter universities and colleges, while the function of adult colleges has become obscure. Especially in China, a country that values diplomas very highly, OUC, with non-academic education as its mission, will clearly face many new problems, and whether the quality guarantee system is complete or not will also be an important factor in shaping its development.

In Chapter 7, Dr. Zhiying Nian analyzes a school credit bank, a new innovation in China's educational reform. Building this school credit bank without a Qualification Framework will definitely entail a lot of ups and downs, but reform by means of a school credit bank has distinctively Chinese characteristics. In China's reform, the experiences of other countries will always be studied to a certain extent. Experts have been heavily involved in the demonstration of Chinese policies. However, in light of distinctive and complicated national conditions, final decisions are often hasty, and in most cases, compromises have to be made. Some compromises are successful in terms of considering national conditions, but undoubtedly some entail sacrificing the core ideas supported by international experience.

Shanghai, as mentioned above, has obtained impressive results in PISA tests, and although this city cannot represent China as a whole, these results can be considered an important achievement for China's educational reform. Note that this is not an achievement of China's "tiger-mom" approach to traditional education, but a new one, obtained by a process of constantly absorbing international experiences and undertaking reform. Shanghai represents a daring vanguard in advancing educational reform. This book is not intended to provide a complete picture of China's educational reform. Rather, it addresses the types of complicated circumstances under which China has made achievements in educational reform, and the conflicts arising in the context of that reform. In this way, this book intends to express the thoughts of some young education researchers.

# 2

# Strongly Promoting the Universalization of Rural Preschool Education to Lay a Foundation for Children's Lifelong Happiness

— Experiences and Insights from Hebei Province, China

**Mingli Fan**

Hebei University, Baoding, China

## Abstract

Rural preschool education development lags behind urban areas to a significant extent in China. In the process of promoting the universalization of rural preschool education, Hebei province has preserved its culture of public welfare, strengthened and fulfilled the leadership responsibility of government, and insisted on "relying mainly on government and collectively established kindergartens, relying mainly on public institution teachers, and relying mainly on government and collective input," the so-called 3mainstays in the development model of preschool education. Consequently, it has expanded the resources of rural preschool education quickly, and more than two-thirds of rural areas have universalized 3-years' preschool education. In this process, Hebei has utilized abundant resources for compulsory education distribution adjustment reasonably and effectively. The Hebei model has been tested in practice and can provide useful experiences and insights for other areas.

**Keywords:** Government dominated, rural preschool education, universalization of preschool education, 3mainstays in the development model of preschool education.

Early childhood education plays an important role in both individual happiness and social development. Recently, many studies in the fields of brain science, psychology, and education have revealed that early childhood is one of the most important periods in the development of emotion, behavior, language, and cognition (Early Head Start, 2004; High/Scope Perry Preschool Study,1993). It has therefore generally been accepted that early childhood education can lay the foundation for one's lifelong development, and can furthermore ensure one's lifelong happiness.

In the process of the development of preschool education in contemporary China, there has been a significant gap between urban and rural areas, and both face respective developmental dilemmas. For instance, urban preschool education faces "the public kindergarten scarcity" and "private kindergarten polarization," with high-quality private preschool education resources becoming too expensive and privileged, while private kindergartens available to the poor are cheap but not well-qualified. Media reports in 2009 and 2010 indicate that "hard to enter kindergartens and too expensive to enter kindergartens" have been a persistent probelme in China. As a result, in order to enable their children to enter a high-quality kindergarten, parents and even grandparents resort to using all their skills and wracking their brains to find the connections needed to enter a kindergarten for the privileged. Some parents or grandparents have spent days and nights in queues to obtain a kindergarten admission quota. Meanwhile, the development of preschool education in vast rural areas lags behind cities, and suffers from an overall shortage of preschool education resources. As a consequence, today rural preschool education has become the shortest board among short boards in the development of education as a whole in China. In 2010, before the promulgation of *The national medium and long term program for education reform and development (2010–2020)*, the Chinese Ministry of Education went to the public for comments, and three modes of soliciting feedback were adopted (phone, email, and letter). Only one problem with preschool education ranked among the top three in this survey; namely, that it is expensive and hard to enter kindergarten. We can say that preschool education is not only an education problem, but also a problem for people's long-term livelihoods and social well-being. Since 2010, the government has continued to strengthen its responsibilities, and has promulgated a series of policies and measures to solve these problems. This study is actually one part of a research project on "the reform and exploration of local preschool education in China." After a survey of preschool education development in 31 provinces, we used a representative sampling method and ultimately focused on 12 provinces as study cases. Hebei province is just one of these cases.

## 2.1 Problems and Reform Opportunities in Rural Preschool Education in Hebei Province

The development of preschool education in a country or a region is closely related to its economic, social and educational development level. Hebei province is located in the North China Plain, and consists of 11 cities and 172 counties, with a population of 71,850,000; 79.5% of this population is rural. Most areas in Hebei province are rural–out of 2,227 towns and sub-districts, 1,969 are in rural areas. Therefore, this study's emphasis on problems facing preschool education development and its 3-years' universalization focuses on rural areas in Hebei.

In geographical location, Hebei province belongs to the East, but at the provincial level, it is a microcosm of China: in rural areas, the imbalance of economic development shares features with Eastern, Central, and Western China. From the geographical scope and the level of economic and social development of Hebei Province, we can say that the area south of Beijing has a features typical of Central China; the area north of Beijing, including Zhangjiakou and Chengde, has features typical of the Western economic and social development approach in China, and in the areas of Qinhuangdao, Tangshan, and Langfang, the features of Eastern areas are more obvious. Therefore, in the study of rural preschool education development in Hebei province, one could say that the case study of Hebei is the sample of the country, with reference to the stages and drivers of early childhood education in rural sub-regions. An analysis of rural preschool education development in Hebei province can therefore inform the study of the stages and drivers of early childhood education in other rural regions.

The stage of preschool education development in Hebei province shows clearly that the development of rural preschool education determines the development trend in the preschool education at the provincial level. On the provincial level, the development of the economy, society, and education in rural areas in Hebei province lags behind that of the urban areas, and the development of rural preschool education is representative of the province as a whole. However, preschool management in the rural areas of Hebei is not standardized, and there are substandard approaches to education, at odds with the leading science on children's physical and mental development.

Therefore, the difficulty with the development of preschool education in Hebei province manifests primarily in the rural areas, and only by realizing 3-years' rural preschool education will there be 3 years' preschool education in the whole province. Recognizing this, Hebei province has taken the correct approach to popularizing 3-years' preschool education by actively promoting the development of rural preschool education.

There are two key points to address in the development and popularization of preschool education: housing kindergartens and recruiting qualified teachers. Since 2000, due to a decline in the school-age population, school layout adjustments offered a golden opportunity for the popularization of preschool education in the countryside of Hebei province. According to statistics, from 2000 to 2006, the number of primary school students in Hebei province decreased from 8,137,300 to 5,095,400, and the number of full-time teachers in primary school decreased from 329,500 to 235,200. In 1998, Hebei province had 45,343 primary schools, and in 2002 had only 28,433, a decrease of 16,910 over four years. The average number of vacant primary school buildings had reached 100 per county, and these school buildings could be converted into standardized kindergartens after renovations and repairs. In 1997, the number of primary school students in Hebei province was 9,011,200, which had dropped to 6,745,500 by 2002; but over the same period, the primary school teaching and administrative staff had grown from 313,000 to 355,300, meaning that rural primary schools were seriously overstaffed. Since 2003, in response to *The decisions on the reform and development for elementary education* (State Council of China, 2001), which proposed a policy to adjust the rural compulsory education schools' layout to suit the local conditions, Hebei province brought about a "win-win" solution by adjusting the layout of reconstruction projects in primary and secondary schools. Beginning in 2005, at the provincial level, Hebei province implemented a "standardization program" for rural junior high schools and primary schools, and according to the standards of construction, used the county as a unit and tailored the layout of school buildings to the school-age population. As a consequence, over 23,000 primary schools would be reduced to 10,000, and middle schools would be reduced from 4,000 to more than 1,500. Thus, the adjustment of the layout of primary and secondary schools was achieved in one step. In the process of eliminating schools with few students and those that were small in size, the province gradually produced a large number of surplus educational resources, including both teachers and school buildings, resolving the problems of lack of funding and teachers for rural preschool education development. This provided a good opportunity for the rapid development of preschool education in rural areas of Hebei province.

## 2.2 Main Measures for the Reform of Rural Preschool Education in Hebei Province

The different areas in Hebei province have the characteristics of Eastern, Central, and Western China. In the process of promoting the 3-years' popularization of preschool education, the province has maintained a policy of offering preschool as public welfare, ensuring that "the development of preschool education is in the vital interests of millions of children's healthy growth and thousands of households, and an important cause for the future of both the state and the nation." During the last ten years, on the basis of its developmental situation, Hebei province has insisted on "relying mainly on government and collectively established kindergartens, relying mainly on public institution teachers, and relying mainly on government and collective input," the 3-mainstay development model of preschool education. As a consequence, it has rapidly expanded resources for rural preschool education, and more than two-thirds of rural areas have universalized 3-years' preschool education. In the process, Hebei province has utilized abundant resources for compulsory education distribution adjustment reasonably and effectively.

### 2.2.1 Clarifying the Public Nature of Preschool Education and Exerting Governmental Authority

In the development of preschool education, Hebei province has always emphasized its public nature, and framed it as the "affairs connected with children's growing-up so as to ensure the family's welfare and the future of the nation" (Pang, 2010). In early years of the 21st century, national policies on preschool education in China have changed, and there has been a tendency in many provinces to sell public and collectively established kindergartens. In contrast, Hebei province has always insisted on its policies of developing preschool education, gradually developing its own typical model of preschool education, especially for rural areas.

In 2001, Hebei province initiated a program "universalizing 3-years' preschool education in more counties and subdistricts," successively using Zhuozhou in Baoding, Shexian in Handan, Shahe in Xingtai, Gaocheng in Shijiazhuang, and Shangyi in Zhangjiakou as pilot areas to evaluate experiences in developing rural preschool education. From 2001 to 2008, there have been five conferences on the fieldwork on developing preschool education in rural areas. In October of 2010, the education department of Hebei held a experience-exchange meeting in Baoding, the theme of which was strengthening the 3-mainstay development model in rural preschool education in Hebei province, and requiring education authorities at each level to improve both the quantity and the quality of preschool education, with a special focus on rural areas. Since 2010, as the problem of "difficult-to-enter kindergartens" emerged in many provinces, Hebei has pursued its own policies in preschool education development, especially the 3-mainstay development model of "relying mainly on government and collectively established kindergartens, relying mainly on public institution teachers, and relying mainly on government and collectivity input." In "Some suggestions on putting great efforts into developing preschool education," which was enacted by the Hebei Government on February 23rd, 2011, there is one special item asserting that the 3-mainstay development model in Hebei should be implemented. This policy also claims that "the governmental responsibility should be fulfilled, and the administrative system and operations of preschool education should be perfected," "accelerating the development of rural preschool education."

Three aspects of the process of promoting the universalization of rural preschool education in Hebei province should be examined closely. First, it makes the universalization of 3-years' preschool education an important objective in the development of elementary education in Hebei, and establishes a developmental plan for the province as a whole and for each city and county; second, it sets down several standards and policies according to the preschool education practice in Hebei, such as administrative methods in kindergarten and preschool classes, in order to guarantee the development of preschool education in the whole province; and third, it brings preschool education into the evaluation of governmental work: from 2003 onward, Hebei province has signed responsibility pledges with its cities for the universalization of 3-years' preschool education. Starting in 2004, the evaluation of the universalization of preschool education has begun in each county, and the government of Hebei province is doing this work step by step.

## 2.2.2 Relying Mainly on Government and Collective Input

As mentioned in the last section, funds and teachers are both important elements for the development of preschool education. One could say that teachers function as the software and funds as the hardware for the development of a kindergarten. As a result, stable and normative financial inputs and diverse and abundant financial guarantees are essential in promoting the universalization of preschool education in rural areas.

In the process of promoting the universalization of rural preschool education, Hebei province insists on "relying mainly on government and collective input." Step by step, the structure of rural preschool education has been established, and it depends primarily on governmental finance, while society and the family also share costs to a reasonable extent. Funds for preschool education are divided into three parts: funds for buildings, funds for teachers and funds for running offices. In the process of promoting the universalization of rural preschool education, Hebei has for the most part used the super-abundant public primary schoolrooms and authorized teachers, meaning that governmental financial inputs dominate the budget. For example, in 2006, total funding for preschool education in Hebei province was 542.18 million RMB, of which governmental finance made up 407.05 million RMB, or 75.08%. If we include the teachers' salary, this proportion maybe even higher (Education Department of Hebei Province, 2007). In 2009, this proportion increased to 76.4% (Newspaper for China Education, 2011).

"Some suggestions for putting great efforts into developing preschool education," which was enacted by the people's government of Hebei province on February 23rd, 2011, emphasizes that preschool education funding should be increased, and orders the governments of both cities and counties to put the financing of preschool education in their budgets. It also demands that more of the new funding should be used in preschool education, and that this proportion needs to be increased over the course of three years.

To ensure that public finance plays an important role in the development of preschool education, all levels of government and their educational administrative departments must also attract funding in a variety of ways while considering local realities. The most popular ways of pooling funds include governmental finance, sponsoring institution input, collective input, input from the masses, individual input, social contributions, and bank loans. Some villages even spend funds for government buildings on preschool education. By 2004, rural kindergartens in Shexian, Shahe, and Wangdu had attained the standard of each school having "three machines, such as a computer, television, and VCD/DVD player" plus "one functioning classroom" (Digest of Chinese Education, 2004). In addition, Lincheng county, which is one of the Key Poverty-Stricken Counties, had invested 7 million by the year 2004, out of which 0.4 million was used to expand public kindergarten (Digest of Chinese Education, 2004).

### 2.2.3 Relying Mainly on Government and Collectively Established Kindergartens

Since the founding of the People's Republic, the developmental strategy in the whole country has involved "placing more emphasis on urban rather than in rural" areas. As a result, although China is developing rapidly, there are many problems, such as the uneven distribution of public educational resources, where high-quality preschool education resources are mostly concentrated in urban areas. After the initiation of education reform and the opening-up of China, demand for kindergarten grew rapidly, and Hebei province insisted on the principle of "relying mainly on government and collectively established kindergartens," especially in its rural areas.

Beginning in 2001, Hebei began to promote the experience of Zhuozhou in Baoding as a model for kindergarten: "three ages, two bases, and five independencies," meaning that kindergarten should enroll 3, 4, and 5-year-old children in different classes, and offer two basic constituents, which are basic kindergarten conditions and the basic hygiene facilities; and "five independence," meaning corporate independence, kindergarten yard independence, personnel independence, finance independence, and administration independence.

Starting in 2004, Hebei began to promote the experience of Shahe in Xingtai, which, rather than insisting on the independence of kindergarten, featured preschool classes attached to primary schools in order to share educational resources with them. The preschool classes here differ from traditional preschools, organizing educational programs under kindergarten rules. Here, the number of kindergartens has grown quickly, especially the government and collectively established kindergartens, which have improved from below-average in quality to above-average for China.

On March 1st, 2011, the "Guidelines for Hebei's Educational Reform and Development in the Mid-Long Term (2010–2020)" put forward that Hebei will carry out experimental reforms in setting up kindergartens, and will place preschool education in the basic public service system of government, paying particular attention to 3-year preschool education in rural areas. In addition, Hebei will maintain the principle of "relying mainly on government and collectively established kindergartens" and will develop experimental, high quality new kindergartens or branches, or take over low quality kindergartens. On February 23rd, 2011, the people's government of Hebei province enacted "Some suggestions for putting great effort into developing preschool education," requiring that all parts of Hebei should make plans for the layout of preschool education depending on their economic conditions, preschool education resources, and public needs. These suggestions emphasize that more public kindergartens should be set up, and that the government and collectives should establish more kindergartens in rural areas. They maintain that rural areas should receive priority in distribution of public funds, and that surplus resources for compulsory education should be used.

## 2.2.4 Relying Mainly on Public Institution Teachers

There are two key elements in the development of preschool education: the guarantee of funding and teacher resources. Today, it is especially important for China to build a contingent of kindergarten teachers with high standards of professional ethics and skills who truly love children. Public institution teachers can better supply this contingent as a more stable workforce that can make progress sustainably. Concerning rural kindergarten teachers, Hebei province insists on the principle of "relying mainly on public institution teachers," thereby consolidating the place of kindergarten teachers in primary teachers' administration. On this basis, Hebei has built a team of kindergarten teachers made up primarily of public institution teachers, with contractually engaged teachers as a supplementary employment force.

In relying mainly on public institution teachers, Hebei province also pays considerable attention on the teachers' quality of teaching, and supplies new teachers through diverse channels. Through the adjustment of compulsory schools' distribution, Hebei captured on an opportunity to place many primary school teachers who majored in preschool education in kindergartens, while placing many surplus primary school teachers into kindergartens after preschool education training. In this process, Hebei did not simply place higher-level school teachers at a lower level, instead choosing those with preschool education backgrounds or interest in preschool education. Through effective training, these teachers learned how to instill good physical and mental discipline in young children and how to educate kindergarten students. Only when they passed a test could they be kindergarten teachers.

Hebei also adopted several approaches to recruiting kindergarten teachers; for example, authorizing teachers who graduated from preschool normal schools, and hiring young people with a high school or higher education diploma, especially those with preschool education or related backgrounds. In order to improve in-service teacher quality, the province has introduced excellent teachers from other areas into local kindergartens, sent local teachers abroad to learn from better educational methods in other areas, trained teachers in teams, and simply trained teachers in their own kindergartens to better familiarize them with the features of kindergarten.

One crucial factor in improving the quality of the profession for kindergarten teachers is to improve their salaries and welfare, which makes authorized positions imperative. According to related national policies, and considering the special situation in Hebei, the provincial government placed authorized kindergarten teachers in primary school positions and made great efforts to enhance their salaries and welfare. For example, in 2003, Shahe in Xingtai introduced authorized positions in kindergarten, and in rural areas the ratio of teachers to children is 1:20. Although still not high enough, this represents great progress. In addition, the salaries of kindergarten teachers are distributed directly by the bank and paid by the Shahe government. In this way, the problem of teachers' salaries not being guaranteed, which has existed for years, has been solved, and the personnel solutions for kindergartens have become more sustainable.

### 2.2.5  Implementing the Popularization of 3-years' Preschool Education and Perfecting Means of Supervision and Evaluation

Starting in 2001, Hebei province began to implement the popularization of 3-years' rural preschool education. They approached this challenge as a battle and fortified their positions, beginning with Baoding Zhuozhou city, Handan Shexian county (a national poverty county), Xingtai Shahe city, Shijiazhuang Gaocheng county and Zhangjiakou Shangyi county (a national poverty county) as pilot projects. In promoting the process, Hebei province held on-site meetings every year in pilot counties and promoted the popularization of 3-year rural preschool education experiences. From 2001 to 2008, on-site meetings were held five times, and in October 2010, the province held a preschool education work experience exchange meeting in Baoding, which clarified the meaning of "relying mainly on government and collectively established kindergartens, relying mainly on public institution teachers, and relying mainly on government and collective input," the 3-mainstay development model for preschool education. This meeting also clarified the requirements of the Education Authorities at all levels to improve preschool education popularization and education quality as their central work, to give priority to government and collective investment, to focus on rural preschool education, and to vigorously promote the standardization construction of kindergartens.

The guarantee to implement the popularization of 3-year preschool education entails qualified supervision and assessment. From the beginning of 2003, Hebei province had signed documents with the municipalities to achieve their targets by managing responsibilities. This launched in 2004, and from 2005 onwards there has been annual balance acceptance and gap evaluation. Starting in 2009, the Educational Department of Hebei province has launched review work on education in the pilot counties to ensure that results continue to be consolidated and improved.

## 2.3  Insights and Policy Suggestions

In China, preschool education, which is an important part of elementary education, plays a fundamental and leading role in advancing individual development in a holistic and healthy way. As preschool education is primarily public in nature, it is the responsibility of a modern government to bring preschool education into the public service system and to govern it effectively (Pang, 2010). Since 2010, the problem of "difficult and expensive-to-enter kindergartens" in China has attracted more and more attention from both the public and the authorities. Accordingly, it is not only a governmental duty to provide preschool education, but also the key to addressing urgent demands and numerous problems with development in contemporary China.

In the process of promoting the universalization of rural preschool education, Hebei province has always upheld the public nature of preschool education and reinforced the responsibility of the government in this area step by step. As a result, the number and size of rural kindergartens has rapidly expanded, and the quality of public and collective resources for preschool education has developed quickly. Today, 2/3 of rural areas in Hebei have universalized 3-years' preschool education. These experiences prove that, in rural areas, clarifying the public nature of preschool education and strengthening the government's relevant responsibilities is very important. They also demonstrate that the development of preschool education cannot be pushed onto the market, especially in underdeveloped rural areas; instead, the government must play an important role.

This study tells us that in the underdeveloped rural areas, only by "relying mainly on government and collectively established kindergartens, relying mainly on public institution teachers, and relying mainly on government and collective input," the 3-mainstay development model for preschool education, can the government guarantee lower fees or offer free pre-school education, and only the guarantee of lower fees or free education can ensure that poor parents send their children to kindergarten, allowing the chain of poverty to be broken, which can ultimately ensure equity in education and for society as a whole. The Hebei model of rural preschool education can provide insights for other areas to draw on, and after studying this model, we put forward three policy suggestions.

First, the development of rural preschool education, especially in underdeveloped areas, should be pursued with public finance, and only on this basis can it be financed in more ways. This requires related departments of government to calculate the costs of kindergartens and lay down standards for financial appropriation and office administration costs per child. On that basis, related departments of the government should introduce preschool education funding into the governmental budget, making sure it occupies a certain proportion and increases annually, and setting up special preschool education standards. We can improve the governmental inputs into preschool education by setting up a mechanism of stable funding for the long-term future in order to make significant progress in practice over time.

Second, the government should be charged with the responsibility of constructing and running kindergartens in rural areas that cannot attract social resources and making efforts to broaden public preschool education resources. In this process, making use of the surplus classrooms and teachers left after the adjustment of compulsory education distribution is obviously one effective approach. Rural areas can also learn from Hebei's experience in rebuilding schools that were in poor condition, while optimizing the compulsory schools' distribution and building standard kindergartens. The school buildings have been repaired and can serve as model kindergartens.

Third, in clarifying kindergarten teachers' status, we should perfect the admittance system for teachers, checking and supervising positions in kindergartens, and gradually constructing ranks for teachers, the majority of whom work in public institutions and are authorized, with engaged teachers as a complimentary workforce. Meanwhile, as there is a surplus of primary teachers compared to kindergarten teachers in China today, we can recruit primary school teachers who have a preschool education background and may be interested in preschool education to work in kindergartens after certain training. Because cultivating kindergarten teachers may cost a great deal and take a long time, it would be better for Hebei to place residual primary school teachers in kindergarten classrooms, depending on their situations, an approach that provide a positive example for other areas to follow.

## References

[1] Chinese Ministry of Education. (2010). *The Speech of Deputy Provincial Governors Long Zhuangwei on the Videophone Conference of National Preschool Education*.
[2] The Digest of Chinese Education. (2004). *Planning and Promoting Macrocosmically to Universalize the 3-Year Preschool Education*. Conference Report.

[3] Chinese State Council. (2001). *The decisions on the reform and development for elementary education.*

[4] Newspaper for China Education. (2011). *To Do the Events Benefitting People Well—How to Resolve the Problem of "Difficult to Enter Kindergartens".*

[5] Education Department of Hebei Province. (2007). *Insist on the Publicity of Preschool Education and Do Efforts to Universalize 3-Year's Preschool Education in Rural Areas.* Report on the National Conference of Rural Preschool Education.

[6] Han, Q. L. (2007). *The Experience of Shahe Established the 2-mainstay Model of Kindergartens and Administrative Systems.* Hebei Province: Lay Foundation for Children's Lifelong Happiness.

[7] Schweinhart, L.J., Barnes, H.V., & D. P. Weikart. (1993). Significant Benefits: The High/Scope Perry Preschool Study through Age 27. In *Monographs of the High/Scope Educational Research Foundation, No. 10.* Ypsilanti, MI: High/Scope Educational Research Foundation.

[8] Pang, L.-J. (2011). To Speed Up the Legislation Progress of Early Childhood Education Act. *Educational Research, 8.*

[9] Pang, L.-J. & Han, X.-Y. (2010). Legislation of China's Preschool Education: Reflection and Progress. *Journal of Beijing Normal University (Social Sciences), 5.*

[10] Pang, L.-J. & Fan, M.-L. (2012). Problems and Challenges in the Management System of Early Childhood Education in Current China. *Research in Educational Development, 4.*

[11] Pang, L.-J., Xia, J., & Zhang, X. (2010). The Free Early Childhood Education Policies of Main Countries and Regions in the World: Features and Inspirations. *Comparative Education Review, 10.*

[12] The People's Government of Hebei Province. (2011). *Guidelines for Hebei's Educational Reform and Development in Mid Long Term (2010–2020).*

[13] The People's Government of Hebei Province. (2011). *Some Suggestions on Putting Great Efforts on Developing Preschool Education.*

[14] Sohu Newsletters. (2010). *"Difficult to enter kindergarten" is not only the affair of a family, but also the affair of a nation.* Retrieved from http://news.sohu.com/s2010/dianji478/

[15] The Statistical Bureau of Hebei Province. (2009). *The Economic Yearbook 2009.* Retrieved from http://www.hetj.gov.cn/extra/col20/2009/010101.htm

[16] The Statistical Bureau of Hebei Province. (n.d.). *The 6th Population Survey Statistical Communiqué*. Retrieved from http://www.hetj.gov.cn/article.htm1?id=4285

[17] The United Nations Educational, Scientific and Cultural Organization. (1996). *Learning: The Treasure within*. Report to UNESCO of the International Commission on Education for the Twenty-first Century. Paris, France: UNESCO.

[18] U. S. Department of Health and Human Services. Administration for Child and Families, Office of Planning, Research and Evaluation, Child Outcomes Research and Evaluation, Administration on Children, Youth and Families, Head Start Bureau. (2004). *Making a Difference in the Lives of Infants and Toddlers and Their Families: The Impact of Early Head Start.* Washington, DC: U.S. Government Printing Office.

[19] Wang, H. (2011). The Intension, Principles and Side-effects of Maintaining the Policy of Government Playing the Leading Role. *Young Children Education (Educational Sciences)*.

[20] Yang, H. (2012). The Equity Theory and Roads to Preschool Education. *Modern Education Management, 3*.

# 3

# Public Primary and Middle School Selection in China: Problems, Reasons, and Governing Policies

## Xiaojie Zeng and Zhiying Nian

Beijing Institute for the Learning Society, Beijing Normal University, Beijing, China

Students prefer better schools and schools prefer better students. Since third grade in primary school, Xiao Ming has spent most of his weekends in training classes: learning math on Saturdays, English on Sunday mornings and composition on Sunday afternoons. He also has piano class on Thursday nights. He has to be outstanding to pass the famous public school selection test when graduating sixth grade in primary school. There is very tough competition in the selection for famous public schools in China. Therefore, gaining entrance to one's school of choice in middle school is called "a war without the smoke of gunpowder."

## Abstract

In China, school selection for both public primary and middle schools is a "bottom-to-top" spontaneous competition among individual parents as a social phenomenon. The study analyzes problems caused by the competition over school selection in terms of money, excellence, and power in public primary and middle schools in China, and explains the reasons behind them. The study then introduces and analyzes governing policies in education.

**Keywords:** China, public primary and middle school, school selection, governing policies.

Today, school selection is undergoing "top-down" reforms in compulsory education in many countries. Charter schools and education vouchers have been widely developed in America. France also canceled the "school card" that prescribed entering schools nearby. In China, on the other hand, school selection among public primary and middle schools is a "bottom-to-top" spontaneous individual competition among parents as a social phenomenon. Gaining entrance to competitive schools using money, excellence, and power has many consequences, resulting in expensive school selection fees, "pre-occupied classes," "students with recommendations from "influential officials," and "students from allied government agencies or enterprises" undermining education equality, introducing a huge burden to parents physically, mentally, and financially. The process also destroys the normal educational order and promotes corruption. Since school selection first began in the 1980s, central and local governments have issued many policies to prohibit the practice. But to date, school selection remains a leading issue in education in China and represents a serious challenge to social equality. This study documents school selection in public primary and middle schools in China and explains the reasons behind the process. The study then introduces and analyzes governing policies in education.

## 3.1 School Selection in Public Primary and Middle Schools in China

China has implemented nine-years of free compulsory education. In order to protect the education rights of school-age children, the compulsory education law of the People's Republic of China ratified in 1986 requires that the local people's governments at various levels ensure that school-age children and adolescents enroll in a school near their place of residence. The Compulsory Education Implementation Detailed Rules, ratified in 1992, reaffirmed this principle. Education equality is the foundation of social equality. The policy of entering a nearby school is intended to ensure the equality of compulsory education. Based on geography, school districts are designated to ensure children have equal educational opportunities. Designating a school district, as opposed to school selection, can prevent students from being divided according to their advantages and prevent the "Matthew Effect[1]" in good

---

[1]The "Matthew effect" denotes the phenomenon that "the rich get richer and the poor get poorer" and can be observed in various different contexts where "rich" and "poor" can take different meanings. The effect takes its name from a line spoken by "the Master" in Jesus' parable of the talents in the biblical Gospel of Matthew.

schools. It can guarantee balance in school and student resources and offer equal developmental space to every school. To promote this principle, the government called off national middle school entrance exams and distributed students among schools using a computer program in 1993.

In order to choose a good school, many parents give up the convenience of entering a nearby school and exhaust all their resources to select a better school. Some good schools offer proportional vacancies for better students by giving exams. As a result, not all schools receive government-distributed students. General schools receive most of the government-distributed students, while better schools receive fewer government-distributed students. Some famous schools even refuse government-distributed students. Take Beijing, for example: according to a survey conducted by the 21st Century Institute of Education in 2011, the proportion of schools receiving government-distributed students in different districts is: 44% in Dongcheng District; 40% in Haidian District; 33% in Xicheng District. Until 2011, seven famous schools in Xicheng District refused to accept government-distributed students. They first opened 10% of their classroom seats to government-distributed students in 2012 ("An Analysis on the Situation of School Selection to Enter Middle Schools from the Scholar's Perspective", 2011). Despite fee-free and exam-free enrollment prescribed by the compulsory education law, there are several approaches besides government distribution to entering school. Many parents describe this fierce school selection process as a "school selection war." The problems with school selection include:

1. Approach of school selection: selecting excellent students.
   The problem: "pre-occupied class."

School selection on the basis of excellence means that good public schools choose excellent students based on their exam scores. This approach is a common, ensures equality of opportunity, and is widely accepted. But in order to select excellent students, some good public schools operate or cooperate with other social training schools to offer exam training classes. They select outstanding students from these training classes and charge a high training fee. Most of these classes are in math, English, and Chinese, and are offered on the weekends. Every famous school has its own corresponding training school. In order to enter famous middle schools, students have to enter their training classes and gain pre-admission by taking several exams. These training classes are what is called the "pre-occupied class." The fierce competition, increasing size, and time requirements of these classes, and correspondingly difficult courses and high fees, all result in problems. Thus the "pre-occupied class" problem has seriously eroded the public education system's ecology.

First, the huge sizes of "pre-occupied classes" and their limits on enrollment intensify the competition for school selection and make entering middle school extremely stressful for students and parents. According to a survey conducted by the 21st Century Institute of Education in 2011, "take Haidian District for example, in 2010, seven famous schools have more than 5000 students, corresponding to 106 pre-occupied classes with 50 students in each class. These schools ultimately enroll 560 students, only 10% of the total number, which means that 90% of students cannot receive a high-quality education in spite of having spent a great deal of time and money" (ibid.).

Second, the long duration of "pre-occupied classes" increases competition for student resources between schools and places an undue burden on students, which affects their physical and mental health. Some students enter a "pre-occupied class" in the third grade while others enter two or even three pre-occupied classes to increase their chances of admission to famous schools. These students get almost no rest on weekends, and the courses they take go far beyond the requirements of primary school. Some famous public schools open experimental classes in advance, asking students in the fifth grade to enroll in order to concentrate excellent students in famous schools. Thus the quality of the student mix in ordinary schools suffers. Under pressure from both parents and schools, this widening gap between famous and ordinary schools seriously undermines the education system's health.

Third, expensive fees place an undue financial burden on parents. Each "pre-occupied classes" costs two to three thousand Yuan per semester. Annual tuition is around 3000 to 5000 Yuan (ibid.). Over the course of four years (from third grade to sixth grade), many parents spend more than one hundred thousand Yuan. School selection thus represents a severe financial burden.

2. Approach of school selection: selecting students who can pay.
   The problem: high and arbitrary charges.

Despite offering nine years of free compulsory education, due to limited educational funds, the government also encourages social organizations to run schools independently and donate funds for schools. In middle school, school selection fees charged by famous schools are considered sponsorship fees, which are recognized by the government and legal. Parents deposit sponsorship fees into a bank assigned by the government or into an education fund account. The government returns a certain proportion of the money for schools to use to improve their facilities and teachers' incomes. For example, the school selection fee stipulated in Beijing is 30,000 Yuan and the proportion returned is 70% to 80%. The proportion returned in Guangzhou is 50% (Zhang, 2011). But some schools break these rules and charge 80 to 100,000 Yuan. The more famous the school is, the higher the fee. On the one hand, there is free compulsory education, while the alternative is an outrageously expensive fee. Some schools even hold back the school selection fee, in cases of corruption. The TV program "Where has school selection fee gone?" in Focus Interview, which was broadcast by CCTV, exposed a phenomenon in which some schools used school selection fees to give teachers bonuses and cover travel expenses or offer other benefits, such as purchasing a new car or redecorating an office. The use of school selection fees falls outside the scope of social supervision. High and arbitrary charges are a cause of widespread concern.

3. Approach of school selection: selecting students with powerful connections. The problem: "students with recommendations from influential officials" and "students from allied government agencies or enterprises."

Some students use political power to gain entrance to a top middle school. Some parents receive the opportunity to send their children to famous schools thanks to the special attention of leaders or the superintendent of the department. They usually meet headmasters in famous schools with a note written by an authority, asking for enrollment. These students are consequently called "note students." In Beijing, banks, state organs, and large-scale enterprises find ways to enroll their constituents' children in famous schools through cooperation with famous schools. They may use public resources or donate supplies (e.g., computers) and funds or they may pay the school fees themselves. According to one survey, the percentage of note students in some schools is 8% to 10% ("An Analysis on the Situation of School Selection to Enter Middle Schools from the Scholar's Perspective", 2011). This approach to using privilege is the most unfair among school selection strategies and plainly infringes on education equality. It is a kind of rent-seeking behavior that affects education and social equality.

## 3.2  Reasons behind the Problems with School Selection in Public Primary and Middle Schools in China

Why is there school selection in public primary and middle schools in China? Why is school selection intensifying in spite of the government ban? The reasons are listed as follows:

1. Huge differences in education quality between schools represent one of the objective institutional reasons for tolerating school selection.

Because of school differences in educational philosophy, quality of teachers and students, school culture and community resources, each school has its own features and schooling level. School differences certainly exist. And there is an institutional reason behind the differences amongst primary and middle schools in China. In the 1950s when the education system was founded, the government had very limited resources and introduced a "key school policy". The Ministry of Education released "Opinions on Focusing on Developing Some Middle Schools and Teachers' Colleges" in 1952. In 1962, the Ministry of Education released "Notice Focusing on Developing a Group of Full-time Middle and Primary Schools". In 1978, the Ministry of Education released a "Tentative Plan on Developing a Group of Key Middle and Primary Schools." In 1983, it released "Opinions on Further Improving the Teaching Quality of General Middle Schools," which still focused on developing key middle schools.

The "key school policy" refers to a means of developing some schools by concentrating finances, outstanding teachers, and other resources on them. China enacted this policy until the middle of 1990s. These hierarchical and differentiated government-dominated schools produce huge differences in education quality between ordinary and key middle and primary schools. The gap between these two kinds of schools applies not only to schooling conditions, but also to teacher quality and morale. Due to the imbalance of educational resources, entering neighborhood schools does not ensure good schooling and excellent education. For the sake of the future development of their children, parents do their best to choose good schools, opting out of neighborhood schooling whenever possible. This is one of the fundamental reasons for school selection.

2. Parents' roles as consumers are another reason for school selection. Today, education rights also include the right to choose a path in education.

In the past, there were key schools and ordinary schools in China, but parents could not choose schools. At that time, it was the key school that had the right to select students, through a form of screening based on exams. Key schools selected students by their scores, a process that was relatively objective and fair. Today, with the neighborhood school policy, exams to enter middle school have been discontinued. Who has access to better schools? In the past, everyone enjoyed an equal opportunity to compete for a good score, which was acceptable to parents. The exam-free neighborhood school policy is seen as unfair by many parents.

Perhaps most importantly, school development must be seen in the context of the establishment and development of a market economic system. Schooling is subject to competition among parent-consumers in a market, and education is no longer a public good, but rather a personal good that satisfies the need for personal development (Dong, 2014). This definitely increases an individual's desire to pursue a better education. The exam-free neighborhood school policy has acted as a kind of catalyst in producing school selection, which in turn worsens the problems resulting from school selection. School selection is therefore an inevitability in China today, and represents a relatively reasonable response to present circumstances. Due to the lack of reasonable institutional arrangements, school selection is intensifying, causing many problems.

3. With a diversified pool of institutions running middle and primary schools, the existence of several different kinds of schools ensures the feasibility of school selection.

In 1985, the "CCP Central Committee Decision on the Reform of the Education System" indicated that local governments should encourage and help state-owned enterprises, social organizations and individuals to run schools and encourage various institutions, organizations, and individuals to make donations for education on a voluntary basis. With these reforms, the appearance and development of private schools represented a break with the single national school system for middle and primary schools. Open enrollment, sponsorship fees and expensive tuition offered a new approach to school selection. In 1992, Shanghai New Century School and Sichuan Dujiangyan Guangya School were built, enrolling students from across the country. At that time, this aroused serious misgivings in educational circles and society. Compared with key public schools that selected students by exam, private schools offered the opportunity to purchase a good education. This is when school selection for middle schools appeared. But at that time, selection was only between public and private schools (Zeng, 1996).

In the 1990s, China promoted further education system reforms and allowed social organizations and Chinese institutions cooperating with foreign counterparts to run schools. The new reforms also permitted several institutions to run a school. Some famous public schools also began to open private schools: for example, starting in 1997, the Beijing Educational Committee decided to separate private middle schools from forty key middle schools. The purpose was to ensure education equality in the period of compulsory education. These middle schools have been transformed into private experimental schools, the so-called "public-running and private-sponsored" and "private-running and public-sponsored" schools that lie between the private and public sectors. In Beijing, the first transformed school was Eleventh School, which began its transformation in 1993. It occupies a leading position in China. Groups of transformed schools first appeared in 1997 ("News Context", 2004). The government permits these transformed schools to enroll selected students by charging school selection fees. Despite the neighborhood school policy in the compulsory education law, these public transformed schools have torn a hole in the system of public education, making school selection a legal behavior to a certain extant. This further encourages the process. Unlike past enrollment practices based on exam scores in key public schools, this is a new school selection system based on a market mechanism.

4. The one-child policy intensifies parents' need to ensure high-quality education and pursue school selection. In order to control a rapidly increasing population, China introduced a one-child policy in 1970.

Each family is permitted only one child, and parents value the education of their children very highly. In China, there is a saying: "Don't let your child lose at the starting line." This is widely recognized by parents who believe that only a good primary school can guarantee access to a good middle school, a good university, and a good job. For future development and access to better jobs, parents are willing to invest a great deal in education. If their neighborhood schools are not excellent, they exhaust every means to send their children to better schools.

The neighborhood school policy in compulsory education was intended to ensure education equality, which is necessary. But when public resources cannot meet the demand for high-quality education, this policy is difficult to enforce, and there is pressure to undermine it even from within the government. Due to weak government supervision and permissive policies, school selection has become not only an educational issue but also a social one. The repercussions of school selection include: violation of education equality, including unequal education opportunities and unfair educational processes and results; school selection with money and power, which means that family background determines education opportunity, hindering social mobility and exacerbating social polarization; and effects on the normal educational order, as school selection promotes "teaching to the test." Famous schools' monopoly on high quality education and student resources also dampens other schools' morale. Finally, school selection imposes a great mental and economic burden on children, affecting their healthy development.

## 3.3 Governing Policies on Public Primary and Middle School Selection in China

At first, the government resolutely banned public primary and middle school selection. But to a certain degree, school selection is reasonable. For this reason, the government changed its policy objective from 1997 to 2010, focusing on controlling arbitrary charges in school selection. To meet parents' demand for school selection, government reforms allowed transformed schools to enroll selected students rather than government-distributed students. But limited high-quality educational resources, strong demand for school selection, inadequate institutional arrangements, and weak supervision all resulted in more and more negative externalities, undermining education and social equality. In addition, the heated school selection competition spread from middle schools to primary schools and kindergartens, causing fierce rivalries and more social problems. The government's current attitude toward school selection has therefore become prohibitive, and a number of governing policies have been issued. In the past, the government launched special projects to manage school selection problems like arbitrary charges, "pre-occupied classes," and the regulation of transformed public schools. Now, the government carries out comprehensive management.

1. Special projects to manage school selection

   a. Policies to manage "arbitrary charges"

School selection fees can solve problems arising from limited education funding, improving school conditions, and allowing famous schools to expand enrollment, providing more chances for students to attain a high-quality education. But without effective supervision, the advantages of school selection fees vanish and distortion in the distribution of educational resources results. Experts point out the increasingly clear disadvantages of school selection fees with regard to the disposition of educational resources, breaking up the original education system and exacerbating imbalances in teaching quality. Unequal educational opportunities widen the gap between different schools over time, negatively affecting educational quality for society as a whole ("Joy and Sorrow of School Selection Fees", 2012). High school selection fees impose a restrictive threshold on receiving a good education and seriously damage education equality.

To control arbitrary and high charges, the State Education Commission released "Suggestions on the Management of Arbitrary Charges in Primary and Middle Schools" in 1995. In 1996 and 1997, the commission repeated these suggestions and put forward a schedule: resolve school selection problems and implement a neighborhood school policy from 1997 to 1998, focusing on arbitrary charges and high charges. In 2000, 2004, 2007, and 2010, the commission also released related documents to regulate and manage arbitrary charges. The Ministry of Education and other ministries co-released a report on "Eight Strategies to Manage Arbitrary Charges in Compulsory Education," forbidding enrollment through charges for open training classes; forbidding charges for the privilege of enrolling in another district; forbidding enrollment through charges for any form of exam; regulating the enrollment of students with special talents; forbidding sponsor fees for admission; forbidding public school enrollment for charges in the name of a private school; enhancing enrollment information and student status management; and increasing enforcement and punishment (Ministry of Education, 2012).

b.  Policy on "pre-occupied classes"

In 2008 and 2009, the Ministry of Education issued related documents on "pre-occupied classes" forbidding public schools and teachers from opening cram schools or cooperating with social educational organizations ("China's Medium- and Long-Term Educational Reform and Development Plan (2010–2020)", 2010).

c.  Policy addressing transformed public schools

Transforming public schools into private schools is one approach to meet demand for school selection. But due to the unclear character of these schools, the informal transformation process, high charges, and even the simple sale of public schools, this approach skews educational tenets and alters the direction of the school system, with serious negative social repercussions. Government documents have prescribed halting approvals of new transformed schools and have imposed new standards on transformed schools since January 1st, 2006. These documents require strict implementation of the regulating transformed-school policy, regulating charges in these schools, and insisting on non-profit principles in education. It was in this year that Beijing began carrying out an overall clean-up and rectification of transformed schools and private schools opened by famous public schools. On November 2nd, 2007, Fengtai District, which includes six transformed schools (half public and half private), released the first reform plan for transformed schools. Three transformed schools in Dongcheng District also turned into public schools in 2008. Today, most of transformed schools in Beijing have been turned into public schools; only a few remain private ("Reform for Schools Owned by Institutions Comes to "an End-year" and Most of them changed to Be Owned by Government", 2009).

2. Comprehensive governing policies

Because of the complex problems at the root of the school selection phenomenon in primary and middle schools in China, the government's special projects have not produced effective results. After 2010, the government began to change its objectives in school selection management, attaching more importance to comprehensive management. Many recently released documents stress balanced development strategies to reduce differences in education quality among schools and thus to fundamentally redress the pressure towards school selection. After the issue of China's Medium- and Long-Term Educational Reform and Development Plan (2010–2020) in 2010, the Ministry of Education proposed ten strategies for dealing with school selection, including a balance of education resources, the transformation of weak schools, the enhancement of supervision, and public consultation ("China's Medium- and Long-Term Educational Reform and Development Plan (2010–2020)", 2010). In 2013, "The Decision on Major Issues Concerning Comprehensively Deepening Reforms" listed school selection management as one task in educational reform and outlined plans to: balance the disposition of compulsory education between urban and rural areas; standardize public schools; introduce a job rotation system for teachers and headmasters; experiment with the school district system; and establish a school district system in which students enter the schools within their respective districts throughout their nine-year compulsory education without taking any examinations.

In January and February 2014, the Ministry of Education released reports and suggestions on furthering an exam-free neighborhood school policy, introducing new requirements for nineteen major cities: first, banning the violation of exam-free enrollment policies and carrying out online registration for enrollment, while enhancing school registry management; second, forbidding competition for student resources and open training classes; and third, banning arbitrary charges for school selection and resolutely investigating and punishing those practices. This new policy requires that by 2015, 90% of middle school students in major cities like Beijing, Tianjin, and Shanghai must enter nearby schools. In each middle school, 90% of students must be enrolled from the school's neighborhood. This proportion should be improved by 2016 and should exceed 95% by 2017 ("Opinions on Implementing the Nearby

Enrollment to Middle Schools Free of Examinations", 2014). In 2014, the government issued many related policies in order to enhance supervision and management, making this year "the most strict enrollment year in compulsory education in history."

In recent years, local governments have also put forward detailed strategies for the comprehensive management of school selection. Take Beijing, for example; the Beijing Municipal Government has adopted the approach of running schools by group for excellent middle schools, following a government-dominated, famous school-led model in which projects are conducted, experts are involved, methods are supported by scientific research, and policies are guaranteed. Famous schools are expected to play a leading role in expanding high-quality education resources for compulsory education. Running schools by group in the six districts of Beijing involves dividing schools into two types of group: one is an education union consisting of several different schools in which one or more famous schools dominate and cooperate with other ordinary schools, sharing excellent education resources, such as the combined high school affiliated to RDFZ (Renmin University of China) and four education groups in Xicheng District (Du, 2012). The famous schools in these education groups offer free resources including teachers, teaching places, and scientific, art, physical, and cultural education resources, and they share courses with ordinary schools.

Consider another example: the education groups of Middle School Four in Beijing, founded in September 2012, built a "cloud classroom" led by Beijing No. 4 Middle School. It offers three networked experimental classes in its member schools (Middle Schools Thirty-Nine, Fifty-Six and One-Hundred and Fifty-Six). These networked classes include a traditional combination of online and offline study and also allow students to choose activities and speeches in education groups. Each Saturday there is layered tutoring for networked classes. Excellent schools in education groups also provide administrative staff and teachers for teaching and management activities, coordination of scientific research, and lesson preparation in member schools. Famous schools also offer training to officials and teachers from member schools, and students in ordinary schools have the opportunity to enter allied famous schools and attend well-known teachers' lectures. For example, the education groups in Middle School Eight in Beijing offer an "overseas study policy" within groups. Beijing Normal University, Capital Normal University, and the Beijing Institute of Education also send experts to participate in networked classes. This communication between famous schools and ordinary schools enhances the education philosophy and teaching quality of ordinary schools by placing famous schools in a leadership role, ensuring mutual development. This approach effectively relieves the pressure of school selection (Wang, 2014).

Another approach to running schools by group involves matching one famous school with several branch schools, encompassing every stage from kindergarten to high school. Students can study in either a nine-year system or a twelve-year system. In either case, the guiding principle is that the middle schools must accept all graduates of the allied primary schools, which greatly relieves the pressure of school selection. There are two approaches to the nine-year system; one is the cooperation of famous middle schools with general primary schools. Newly enrolled first-grade students in primary schools directly enter the corresponding excellent middle schools. Qingnianhu Primary School in Dongcheng District and middle school One-Hundred and Seventy-One operate in this mode. The second approach to the nine-year system is to extend the education stage in excellent primary schools by three more years, adding middle schools to cover the nine-years' compulsory education ("Beijing Plans to Establish 7 Nine-year Education Schools from 2014", 2014).

There are currently 93 schools operating with a nine-year system. In Beijing, the two modes of nine-year system are "six plus three" and "five plus four." For instance, Jingshan Middle School is a "five-four-three" school system, offering five years of primary school and four years of middle school. Other schools like Middle School One recombine courses in middle and primary schools, listing computer science and English as basic courses starting in the first grade. Middle School One also adopts a sliding scale system to enrollment, dividing fifth-grade students into two groups: one entering sixth grade and the other entering first grade in middle school in advance (An, 2014).

In addition to promoting a better balance of educational resources and relieving the pressure of school selection by offering more excellent resources, the Beijing municipal government has issued a series of new policies to fight privilege, prohibiting the practice of enrolling "students from allied government agencies or enterprises" that had lasted for years. Beijing now forbids "selective enrollment" (enrolling excellent students by exams in advance) and "pre-occupied classes." Meanwhile, the computerized enrollment system offering a unified admissions service for primary and middle school is being used to enhance admissions monitoring. Every famous school is required to declare an admissions plan. The entire process of entering middle school from primary school has been piloted in Dongcheng District for public demonstration and its results were published for online enquiry on June 13th.

The Beijing municipal government has also introduced open information on entering middle school from primary school, ensuring fair and transparent "sunshine enrollment."

There are eight thousand five-hundred and forty-six students entering middle school this year in Xicheng District in Beijing, and only four thousand three-hundred and forty-three students will participate in computerized distribution, accounting for 50.8% of the total number (this proportion is only 33% in 2011). The proportion for all of Beijing is 76.92%, and the proportion of entering primary school from kindergarten in Beijing is 93.7% ("Beijing Municipal Commission of Education: to relieve the fever of school selection through more high quality", 2014), relieving much of the heated competition for school selection. It has been predicted that with the increasing proportion of enrollment handled by computerized distribution, the government's promotion of school development, and the normalization of entering neighborhood schools, school selection problems will, in fact, be solved. School selection will not resurface and define the nature of education – there are to be no social distinctions in teaching.

## 3.4  Conclusion

In China, the nature of school selection competition in public primary and middle schools differs from that in Europe, America, and other countries. School selection in Europe and America deliberately introduces a competitive mechanism to improve the school system, while in China, it is a spontaneous behavior and a poorly regulated social phenomenon. In Europe and America, school selection is promoted by the government and stresses performance, whereas in China, school selection is prohibited, and the government stresses equality. First of all, the intensity of school selection pressure in China reveals a contradiction between parents' strong desire for excellent education resources and an inadequate overall supply of education resources, which is a very common type of social problem in China. Second, with the establishment of a market economy, the idea that education is a consumer good with an individualistic character and that education rights include the right to select a specific path in education are widely accepted by parents. School selection, in this sense, represents a kind of awareness of these rights. Third, in China, because of the increasing variety of parents' educational needs and educational choices, it is common for students to choose private schools, study abroad, or even study at home. Therefore, school selection in public primary and middle schools must be considered an important policy issue that needs to be resolved.

Fourth, despite the government's policy of "forbidding school selection," merely prohibiting the behavior without resolving its underlying causes will only exacerbate related problems. Fifth, from special projects like canceling key schools and arbitrary charges in school selection to comprehensive management for balancing education resources, the recent changes in education policy also promote institutional reforms such as teacher flow mechanisms, the running of famous schools by groups and the nine-year educational system. Thus, in the process of addressing the problem of school selection, public primary and middle schools in China are undergoing broader forms of reconstruction with wide-ranging effects on quality and equality in education.

## References

[1] An Analysis on the Situation of School Selection to Enter Middle Schools from the Scholar's Perspective: school selection with money and power. [N]. (2011, September 5). *The Economic Observer*.

[2] An, S. (2014, April 17). The Amount of Nine-year Education Schools in Beijing Reached to 93. [N]. *Beijing Youth Daily*. Retrieved from http://edu.sina.com.cn/zxx/2014-04-17/0958415480.shtml.

[3] Beijing Municipal Commission of Education: to relieve the fever of school selection through more high quality. [N]. (2014, July 29). *The Beijing News*.

[4] Beijing Plans to Establish 7 Nine-year Education Schools from 2014. [EB/OL]. (2014). Retrieved from http://learning.sohu.com/20140116/n393614387.shtml.2014-01-16.

[5] China's Medium- and Long-Term Educational Reform and Development Plan (2010–2020). (2010). Retrieved from http://news.163.com/10/0729/20/6CPN8B1M000146BC.html

[6] Dong, H. (2014). To Cool Down the Fever of School Selection: from "inner-governance" to "social governance" [J]. *Global Education, 7*, 50-62.

[7] Du, D. (2012, October 26). Beijing Key Middle Schools Launch the Model of Running Schools as Educational Group to Eliminate the Pressure from School Selection [N]. *The Beijing News*.

[8] The Educational Group formed with 4 Key Middle Schools including Beijing No.4 High School Absorbs the Sharing Resources among 15 Common Schools. [EB/OL]. (2012). Retrieved from http://news.cntv.cn/China/20120912/101989.shtml.2014-08-08.

[9] Joy and Sorrow of School Selection Fees [N]. (2012, October 19). *Citizen Herald*.

[10] Ministry of Education. (2012). Eight Measures to Rectify Arbitrary Charges of School Selection in Compulsory Education. Retrieved from http://news.xinhuanet.com/edu/2012-02/24/c_122747102.htm

[11] News Context: From Computer Distribution to Institutional Transformation of Public Schools [EB/OL]. (2004). Retrieved from http://news.sina.com.cn/o/2004-09-20/10253716167s.shtml

[12] Opinions on Implementing the Nearby Enrollment to Middle Schools free of examinations. (2014). Retrieved from http://news.sohu.com/201401.

[13] Reform for Schools Owned by Institutions Comes to "an End-year" and Most of them changed to Be Owned by Government. [EB/OL]. (2009, May 8). *China.com.cn/news*.

[14] Wang, B. (2014, April 10). Two-year Established West District Educational Group Promote the Balanced Development of Education to Shape Qualified School Groups [N]. *Morning Post*. Retrieved from http://bjcb.morningpost.com.cn/html/2014-04/10/content_281004.htm.

[15] Zeng, X. (1996). The Analysis and Suggestions on School Selection in China [A]. "School Selection System" and Compulsory Education Reform in America [D]. Beijing: Beijing Normal University.

[16] Zhang, L. (2014, August 8). Intractable School Selection Fees [N]. *Chinese Economic Weekly*, *40*. Retrieved from http://business.sohu.com/20111017/n322477456.shtml.2014-08-08.

# 4

# The Era of Governance: The Reform of Regional Education with Municipal Co-ordination

**Yijin Zhang**

Faculty of Education, Beijing Normal University, Beijing, China

As a prefecture-level city in Eastern China with a population of ten million in Shandong Province, Weifang continuously contributes to the reform experience of the entire country. In recent years, education reform in Weifang has attracted groups of experts and officials at all levels of education. They want to know why Weifang, as a prefecture-level city, has managed to promote education reform for more than ten years, and can solve difficult problems in so-called "deep-water" areas, such as school selection in the city, the heavy burden on primary and middle school students, and the imbalance in education between urban and rural areas. Another component of this issue is the value and the role that prefecture-level educational governance, as a medium-level organ of government, has to play in the whole education system, from the central to the local.

## Abstract

In education reform that is "governance"-oriented, an important issue and task is determining how to implement "decentralization" and establish a modern education governance system with accountability for performance. The Ministry of Education has clearly made the decision to decentralize and expand provincial education and manpower, but after decentralization to the province level, the issue of decentralization has arisen once again. Once provincial-level co-ordination is realized, what power still needs to be decentralized? Where should it be placed, and to what extent? What goal is to be achieved, and how? There are no experiences to refer to in order to answer these questions. Taking Shandong as an example, this chapter discusses the role administration plays in promoting the comprehensive reform of education. This is the only way to establish a governance system in modern education and to improve the capabilities of education governance.

**Keywords:** Education governance, municipal co-ordination, decentralization, regional education, Weifang.

In November 2013, for the first time, the Third Plenary Session of the 18th CPC Central Committee put forward a plan "to push on with the modernization of the country's governing system and capabilities," which has widely been interpreted as meaning that China is beginning to enter "the era of governance" ("The CPC Central Committee's decision on a number of important issues of deepening reform and opening up in an all-around way", 2013). "Governance" is a richer and more inclusive concept than the traditional approach to "management," with an emphasis on managing multiple subjects, and on democratic, participatory, and interactive management, rather than a top-down approach ("How can Governance System and Governance Capacity Realize Modernization", 2013). In the current system of educational administration, a central authority is responsible for macro top-level strategic design, and the provincial level takes on the important task of implementing co-ordination, while the county assumes the main responsibilities for compulsory education management. Among these responsibilities, only that of municipal education is unclear. As a result, people tend to regard the Municipal Bureau of Education role as relatively unimportant, consisting mainly of conveying documents and reporting issues. But the Weifang City Bureau of Education is out of the ordinary, having planned and coordinated education reform for 17 counties (cities) and made an innovative breakthrough in key difficult areas, providing the national education system of governance with a striking regional example.

## 4.1  How Much Can a Municipal Bureau of Education Do?

What can a prefecture-level City Bureau of Education do? The subtext of this question is whether a Municipal Bureau of Education is that important. It seems that the central authority and the province are in charge of education top-level strategic design and co-ordination, while compulsory education is county-centered, and in the current education administrative system there is no important role for the Municipal Bureau of Education to play, other than conveying documents and passing information from the top down. As the

Municipal Bureau of Education has less power, it seems unnecessary for it to take on more responsibility.

On the other hand, as the affiliated department of local government, the Bureau of Education exercises authority on behalf of the government. According to this study, this kind of administrative habit – "only up" (caring only about the superiors) – brought about many unfavorable consequences. For example, working is liable to turn into a mere formality, making it difficult to attain a good result, while the staff members are too busy and easily ignore whether the core functions of the Bureau are performed. There are all kinds of local educational administrative organizations with too many staff members, leaving schools "overburdened." The local education administration, in turn, is required to implement new policies. In the process of performing their functions, local educational administrative organizations are much more concerned with demonstrating "what they've done" while ignoring "how well they did it" (Ling, 2010).

Weifang City's Bureau of Education is different: from the staff to the chief and the secretary, from the secondary sector to the core bureau, daily work is no longer a matter of being "at the office." Instead, everyone is involved in specific reform projects and is doing their best. This has all come about through recent reforms. More than ten years ago, the Weifang Education Bureau didn't have this many responsibilities, and working pressure was nevertheless higher than it is now.

At that time, parents often complained to the City Bureau of Education about heavy schoolwork, and some people even wrote things to the secretary like: "Look at the street in the morning – besides the sanitation workers, the students and their parents are the first ones to leave home. The children don't even get enough sleep. This is heartbreaking!"

At that time, the City Bureau of Education became a reception room during the teachers' title evaluation period, when officials came to consult the policy, when parents came looking for preferential treatment, or when someone came to complain. The Bureau was almost overwhelmed with catering to the needs of all kinds of people. Then there were 935 cadres above the vice section and many "branch-level principals" and "division level principals," while there were few principals as educators. When some county Bureaus of Education held a meeting, the principal of County No.1 Middle School even didn't attend the meeting in person, thinking too highly of his own position to be bothered. At the same time, the City Bureau of Education was also promoting curriculum reform with documents issued one after another, calling for "providing comprehensive courses" and "attaching great importance to students' comprehensive abilities," but after several years the regional education ecology had not changed, the schoolwork was still heavy, and people remained dissatisfied with education.

The City Bureau of Education held a meeting to rethink the causes of these problems: *Didn't our educational administrative departments work hard?* No. From the Bureau leaders to the heads of department, everyone united to reform, and worked "five plus two," which means all week and "day and night." *Didn't we concentrate on working?* No. We made an effort to promote curriculum reform by printing 30,000 copies of ten regulations for running primary and secondary schools, which were posted to every classroom in every school of the city. Meanwhile, we made requests at every meeting, and we made a greater effort to inspect, assess, and urge. *But why weren't there improvements? Why was the Department of Education still to blame? What was the root cause of these many repeated and entrenched problems?* Upon reflection, there seemed to be only one problem: what could the City Bureau of Education do if the education system wasn't changed?

If the senior high school entrance examination didn't change, the students would still have to do a great deal of homework for teachers. If the school didn't eliminate the administration, the principal would still pursue an upper official position, and no one would be willing to go to the schools in remote rural areas. If the power was still concentrated in the administrative department, the construction of the modern school system would be in vain. Every time these problems arose, they seemed to represent barriers to progress. *Should we avoid the issues or resolve them?*

Guohua Zhang, the chief of Weifang City Bureau of Education, said, "The education administrative departments take on their own responsibilities, so any department will become a serious obstacle to the grassroots reform if it doesn't take on its responsibilities. The city's duties include not just supervising the counties, monitoring implementation, and evaluating, but also coordinating and managing schools, principals, teachers, and quality education entrance examinations and other core elements. We should make the responsibilities of the cities clear and build a city-based regional educational integration operating mechanism and linkage mechanism to co-ordinate the development of education."

## 4.2  Education Reform Should Begin with the Functional Reform of a City's Bureau of Education

With executive administration leading the way whenever an education reform program is put forward, people have no feasible outlet through which to convey their demands. The disparate functions of education departments only serve the needs of higher authorities, so the problem is difficult to solve even with considerable reflection. But these problems represent key obstacles, blocking the development of education by piling up over time.

"Too many factors are involved in education reform. In the past, we got used to sitting in the office and passing on documents and holding meetings. It now appears that such an approach to management has failed." "Only reducing the burden on students or only reforming the curriculum won't work. Reform should be comprehensive." This sharp self-criticism is from a meeting that the leadership of the Weifang Bureau of Education held in 2008. This meeting led to a great change in the functions of the Bureau of Education, establishing a comprehensive education service platform that allows the administrative staff and the people to communicate face to face.

Soon, the "Weifang City Education Service Center for People" came into being. The Bureau of Education made a commitment to the public through the newspaper, radio, and television: all feedback and needs related to education can be brought to this "one-stop" service hall to seek help and solve problems. In order to provide convenience, eight sub-centers were set up, including social training centers, a school-enterprise cooperation, alumni development resources, family education, student financial aid services, counseling complaints units, normal students' employment services, and study abroad centers. All these services, which closely related to the public's interests, were separated from the Department of Education. Nearly half of the staff in the Bureau of Education moved to the center for serving people, comprising 21 staff members from eight departments including compulsory education, private education, organization and personnel departments, and four subordinate units.

Every day, the service center was overwhelmed by incoming calls about people's educational needs, opinions, and suggestions. For simple consulting, the staff were able to reply on the spot, and matters such as approval were handled instantly. But there were still some questions that the staff couldn't answer, especially as related to inter-departmental management and institutional problems. Which would be better: to reply in a routine way or to find the solution by going outside of that routine? The leader's attitude was clear: issues related to people are important, and no matter what is involved, the Department of Education would solve their problems.

The leader of the Bureau of Education soon regulated the work process, including, for example, the way the "front desk and back office" coordinated on problems that couldn't be solved on the spot, which would then be transferred to relevant departments. Particularly serious problems would be directly referred to the main leader and dealt with immediately. The Bureau introduced "daily news," "weekly reports," and "monthly analysis reports," and the problems about which people were highly concerned would be selected and then submitted to the Bureau of Education leader's office. "Project management" was carried out for serious problems that couldn't be solved by any single department, which entailed a bid to collaborate with well-known experts at home and abroad to find a breakthrough. Since the project launched, each project has had its own budget, by way of "doing what needs to get done and then taking the money." Projects would be evaluated by an assessment team from the Bureau of Education in the end. The yearly evaluation is no longer provided by the departments, but instead is based on the project.

Like an enormous octopus, the service center not only has antennas reaching out to thousands of families, but also directly transmits information on their demands to the central system of the Bureau of Education. Often, all administrators will take action, from the service center to the City Bureau of Education and the departments of the Bureau in response to just one call from the public, even launching a project to solve the problem with multiple parties. The municipal Bureau of Education now has more than 50 staff members and has built up a platform through which to communicate with tens of millions of people, trying its best to satisfy every demand.

Although the service center has earned much praise in Weifang, the education policymakers assume that this way of "waiting" to solve problems is still "not perfect." It wants officers to leave the office, enter the community and the schools, and identify and solve education problems, thereby making every education administrator directly responsible for reform.

In 2009, the City Bureau of Education organized all government members into a supervision team, and each department has an area of responsibility. The more often the staff of the Bureau of Education contact individual members of the public and schools, the more information and detail they will take into consideration. Jiao Tianmin, the chief of the Organization Personnel Section, said, "In the past, what we did depended on the interests of the department, but now what everyone does is part of the reform process."

Problem-based reform removed the barriers between departments and established the concept that the whole was greater than its parts, so that work could be done by concentrating on important and difficult problems. The establishment of these areas of responsibility gave each administrator a platform from which to play his role, along with a before-and-after evaluation mechanism. In this approach, no serious problem was referred to others; instead, problems were solved together.

## 4.3 Give the Power to the Third Party

The more widely the service center's door is opened, the more education problems are exposed. The number of calls coming in can be compared to raindrops: a case of corporal punishment in a village school; a kid in the town middle school starving because he didn't pay for his meals in time; children being made to take medicine in a private kindergarten; etc. Inspectors also often encounter such problems, as what the inspector has seen is a qualified curriculum schedule, while another class schedule is running again after the inspection, etc.

Blowout problems challenge the educational reformers in Weifang. There are only 50 staff in the Bureau of Education, serving more than 4,000 educational institutions, 1.5 million students, and more than 300 million parents. How can a small-scale department serve this population? And how can its limited human resources meet the needs of millions of people?

The City Bureau of Education decided to make advantage of outside forces and invite third parties to do the things that the City Bureau of Education was unable to. In November of 2008, the Innovative Education Management and Assessment Center, an independent third-party organization in Weifang, was established. This affiliated supervision body signed a contract with the Bureau of Education to undertake routine supervision tasks and deal with complaints. At the end of the year, the supervision body receives corresponding payment after being evaluated by the Bureau of Education and representatives of the parents. The supervision body consists of retired principals and old inspectors responsible for visiting more than 3,000 schools (kindergartens) in 119 counties and 12 towns (cities), coping with more than 5,200 complaints. Thanks to these efforts, the supervision body found that parents complained less often about irregularities, and were more concerned about issues beyond the scope of education.

The supporting efforts of this supervision body changed the power distribution pattern in Weifang's education management. In the past, the Bureau of Education played both the role of athlete and referee, so to speak, as they wanted to run everything, but ended up with dissatisfying results. Now other social forces take part in supervision and have veto power in evaluations, and as a result, all those previously insoluble problems have been successfully solved.

The power of third parties is not limited to supervision, but extends to taking part in the city's educational decision making. In 2013, many schools reported that the heavy burden of coursework in basic units was a consequence of holding too many appraising and commendation activities at the end of the year, recommending that these should be decreased. The Weifang Institute for Innovative Education Policy was entrusted with the review of this proposal, and 11 principals and representatives of teachers evaluated 55 commendation items one by one, ultimately keeping only 11 items.

This "bold decision" on the part of the Weifang City Bureau of Education surprised many people: it involved empowering a third party to arrange the hearing of all stakeholders before decision making, permitting the third party to organize experts participating in the process of supervision, and having the third party evaluate the proposal's efficiency independently after supervision. For example, the documents issued by the City Bureau of Education were to be assessed by the experts chosen by the third party. All matters related to the interests of students, teachers, and parents were to go through third-party hearing procedures, whether they were rules and regulations or policies and documents transmitted from the top. Project management, senior principal selection, and teacher employment have been the focus of third-party evaluation. There are now six teacher-training centers in charge of teacher training, which can be chosen freely by teachers with "training vouchers."

In fact, this move established a democratic consultation mechanism for transparent decision making. Studies have shown that the driving concept behind China's public governance decision consultation on education is democratization, which also has its own logic and meaning: what kind of boundaries and freedoms are appropriate in democratic education, and how does this system of education avoid sinking into absolute anarchism and liberalism, and thus also avoid falling into the trap of pursuing self-interest in the guise of the public good? This requires stipulating the boundaries of democracy and freedom in education governance, from the perspectives of the government, non-government organizations, individuals, and others (Tao, 2010). Now deeply involved, "the third party" has taken over some of the power of the Weifang Education Bureau and created a democratic consultation, producing consensus and a win-win situation in which it is no longer the Bureau of Education that makes final decisions. Starting with institutional reform, the transformation of functions and decentralization, the Weifang Department of Education consistently broke through barriers and innovated, spending several years forming a complete set of rapid response mechanisms with which to discover, research, and solve problems.

## 4.4 The Reform of the "Deep-Water Area" Is Underway

When the City Bureau of Education began to design its new education scheme, there was no comparable experience from which to learn in the whole country.

The policy of the City Bureau of Education is that as long as it is beneficial to children's healthy growth and promoting the quality of education, anything inappropriate must be changed.

In the past, although a comprehensive curriculum was required, non-examination subjects were often replaced by main subjects. 12 subjects are included in the new senior high school entrance examination, while, music, art, and IT courses are also offered. In the past, these tests had life-long consequences for children's potential. The new senior high school entrance examination uses repeated examinations, so that starting in the first grade in junior high school, students can choose to take part in different subject examinations at different times. Students cannot only take the examination in advance, but can also re-take the examination to achieve a better score. In the past, high school admissions were closely tied to these results, enrolling students from key middle schools first and ordinary middle schools second. The new senior high school entrance examination gives every school the power to set its own admissions standard, so that children with different qualifications and interests can still find a suitable school.

Even so, the examination is related to millions of students' interests, and linked to thousands of families. Although the municipal Bureau of Education had the authority to co-ordinate the exam, they chose not to implement this reform all at once. They let the final decision be made at a meeting in the mayor's office. The mayor convened the leaders of relevant departments to study the new scheme word for word, and it was ultimately unanimously adopted, with an emphasis on the fact that the departments must provide protection for examination reform. But even the government's support was not enough to implement these reforms: the Bureau also needed to see whether people approved of them. So the city issued a special edition in the local newspaper with the greatest circulation, the "Weifang Evening News," discussing the new examination in detail. Meanwhile, several hotlines were opened, waiting for parents' complaints or feedback. One day, two days, and then a week later, there were no calls; everyone was surprisingly calm. At that time, polls showed that 96.9 percent of parents were (very) satisfied and 92.3 percent of students were also (very) satisfied.

The Bureau of Education then realized that people had been looking forward to such a reform. The reform showed respect to the public, so they supported it. Starting in 2004 with a small-scale test, then in 2006 with comprehensive reform, and finally with continuous improvement this year, examination reform in Weifang has continued. It has won public approval, but education reform in Weifang will not stop here. To improve the education ecology radically, there are many difficult problems left to solve and many barriers left to break. Consider the following:

- After the senior high school entrance examination, in order to give students more choices and better opportunities for further development, Weifang made great efforts to develop and expand vocational education, and removed the barrier between high schools and vocational high schools, so that students in vocational high schools and students in high schools would have the same opportunity to go to college.
- The school has become the major driving force of reform. If the principals do not really understand education, no reform scheme can be carried out. In 2004, Weifang began to reform the official rank of principals, and established four levels (junior, intermediate, senior, and special) with nine grades in a performance-oriented assessment system. The authority to appoint school principals belongs to the Bureau of Education. Since then, there have been no opportunities for administrative cadres to take on this job just for promotion. A large number of education experts have become the main force behind reforms.
- The key to the modern school system is to give the school the power of decision making. The City Bureau of Education returned enrollment autonomy, the right to hire middle-level cadres and staff, the right to professional title evaluation, the right to use funding, the right to distribute the staff's income, and the right of evaluation of the schools, entitling the school to be the one taking responsibility for reform.
- If there are only a few high-quality public schools, parents will still desperately compete to select a better school. Weifang put forward a mixed development system for private education: "Society invests to build the school; the government provides teachers; the school charges to ensure the work; the departments are coordinated to supervise," greatly strengthening education. Currently, to choose a school, you can also choose a private school.
- If the education officials don't understand education well, they are likely to give the wrong commands and even misdirect education reform. The City Bureau of Education coordinated personnel departments and standardized the leader's appointment of the Bureau of Education in the county (city): the secretary must have experience in teaching or be

a graduate from a teachers college. At least two-thirds of the leaders have professional experiences in education, and some have served as principals.

It is only the courage to break through that can bring about continuous improvements. Every breakthrough in the education reform in Weifang begins with solving specific problems. Every breakthrough is about making education more "relaxed," so that the whole region's education system begins to radiate vitality.

## 4.5 What Will Education Be Like after "Unbound"?

What will the school be like after "unbound"? How will the principals, teachers and students change?

"In the past, the secretary of the City Bureau of Education held the power to evaluate the principal, so the principal's work centered closely around the secretary. Now, when the assessment of star schools begins, the scores account for only 30%, while the specifications for running a school, curriculum implementation, the overall quality of students, and the satisfaction of the public and the parents account for 70%. The principal's work focuses closely on issues about the teachers, students, and parents, and all his energy is concentrated on education." Yunfeng Yu, principal of Weifang Middle School, explained how the focus of the principal's work has changed with the shift in assessment orientation. Principal Yu attended class everyday to learn about the status and needs of teachers and students. He made every effort to promote multi-level teaching, carrying out selective courses in nearly 200 subjects, such as making pottery, carving, and kite-making, which have been well-received by students.

The principals' rank assessment system, in four levels (junior, intermediate, senior, and special) with nine grades, leads principals down the path of performance-focused and teaching-centered experts' growth. The Ministry of Finance for the city and the county budgets 16 million for the principals' wages. According to their performance appraisals, the Departments of Education pay "junior, intermediate, senior, and special" principals 25%, 35%, 40%, and 80%, respectively.

The implementation this ranking system makes the profession of principal a public career, and there are 1016 candidates for each job. The school is no longer a place for "officials," and the principals don't select a school for its administrative level any longer, instead focusing on where they can realize their own ideals and values. Holding on to this idea, Junqing Zhou went to serve as the principal at Yaoge Village School in Gaomi City. In 2011, when he arrived at the primary school located in the urban-rural fringe area, student attrition was a serious problem. By boosting the morale of the teachers, improving management and the school's environment, seeking the support of the Village Party, and so on, Yaoge Village School succeeded in a "counter-attack." The number of the students increased from 598 in the first year to more than 1,100. In 2012, Yaoge Village School won first place in the comprehensive monitoring and evaluation of the city's education.

The data shows that since the principal ranking system was implemented, 469 principals have gone from the city to the village, from high-quality schools to poor-quality schools. The "reverse flow" of principals is promoted in the city fringe area to let highly skilled principals renew poor schools.

The rules for evaluating principals have changed, and so have the rules for evaluating teachers. Taking the teachers' professional title evaluation as an example, schools now have the right to evaluate teachers, and teachers have the right to design review schemes that will be implemented as long as more than 85% of teachers adopt them. The selection criteria do not focus on papers and open classes, but rather on the effectiveness of classes and the feedback of parents.

With no interference from external actors, the principals and teachers finally returned to class. Kong Juan, a teacher in the Tenth High School of Weifang, said that without the pressure of entrance examinations, teachers spent more time on classes, prepared lessons together, and shared "listening tidbits" with other teachers and students. The teachers in Guangwen Middle School were more careful. They analyzed each student's learning situation, including different abilities in different subjects, social responsibility, degree of anxiety, expectations, relationships, hobbies, etc. Based on this holistic assessment, the students were grouped into individualized classes.

The proposition concept and the way the contents of the senior high school entrance examination have changed have freed students from intensive exercises. From time to time, subjects closely related to life and politics appear in the test, so teachers also introduce students to real life situations. In 2009 and in 2012, the Ministry of Education's curriculum center surveyed 47,000 primary and secondary students in 590 schools in Weifang. The findings showed that the main indicators of academic burden and sleep time, which reflected students' academic status and the quality of education, were significantly better than the national average. This addressed pervasive problems in Chinese compulsory education, such as lack of sleep, heavy academic work loads, and excessive anxiety.

In 2012 and in 2013, the Provincial Education Board in Shandong province entrusted a public opinion survey center with monitoring students' academic workload in the province. The results showed that students' academic workload in Weifang was the lightest among 17 cities.

## 4.6 The Key Is Improving the Environment in Education Management

In recent years, education reform in Weifang has attracted batches of "learners" from all over the country. Someone said that it was lucky that the Bureau of Education in Weifang enjoyed the concern and support of the local party committee, the government and the community, which were beneficial to the reform. Without this support, the educational administrative departments couldn't make these breakthroughs.

Taking the principal official ranking system as an example, four party secretaries, three mayors and two directors of Weifang made great efforts to introduce this system since 2004. Nineteen types of red tape were issued in eight years, and the personnel, finance, and preparation departments closely cooperated with each other to support the policy, finally implementing the ranking system successfully.

Guohua Zhang, secretary of the Bureau of Education in Weifang, believes that a good reform environment is not formed naturally, but depends on the education departments' capabilities. The education departments should be good at seizing opportunities and promoting educational reform and development, and should get the major stakeholders to plan and design education reform together with development of the economy and society. Based on these fundamental values, they not only won "preference" from local party committees and governments, but also fundamentally straightened out their complicated relationship with education.

For example, they clarified the relationship between the Bureau of Education and the schools to make the principals work on school issues actively. They also clarified the relationship between education and society to meet educational needs at all levels in society; between the school and the community, while the "immediate boss" of the school changed from the Bureau of Education to the council; and amongst the educational administrative departments, while "small sectors" undertook "great service."

Someone wondered aloud whether education reform would succeed depending on "power," which determines interests. When the Bureau of Education handles the relationships among these groups, decentralization and alienation are always involved. If you are careless, things will not turn out the way you want. In fact, the core of decentralization in Weifang education is not a matter of relinquishing authority, but rather one of changing operating modes to distribute power more reasonably so that every stakeholder in reform can find their own position, and make their power boundaries and scope of responsibility clear. They carry out their duties and demonstrate their ability to follow a "multi-center, multi-subject" career development pattern. This is a problem that needs exploring and solving in the governance era, especially in pursuit of promoting education governance capacity and a modern governance system.

Only after straightening out the various relationships was a good regional education governance environment formed, after which the education and the market could interact positively. This confirmed education policy researchers' observation, that "the relative autonomy that education has is rooted in a particular way of achieving socio-economic reproduction". [5] Now in Weifang, no one regards education as a department with "only money," but rather as the "engine" of the city's economic and social development.

As a manufacturing base, human resources are a lifeline for the economy in Weifang. In August 2012, an order valued at three billion excited a Weifang company, but due to a "labor shortage," the company was unable to take on the "big order." The business owner was in a desperate situation, and several government departments made an effort to help, but the effort ultimately failed. Ultimately, it was the City Bureau of Education that resolved the crisis: a workplace experience program lasting three months was introduced in the production line, and 4,176 students from 15 vocational schools combined learning with an internship at the plant, which resolved the crisis in time. Their education contributed to the local economy's development, and in turn, the enterprise supported education. The market and the education system can do nothing without each other.

After continuous education reform had been promoted for more than ten years, Weifang became well known as a place for learning, with many elite schools and famous principals and teachers, leading to more and more public satisfaction with the education system. Starting in 2008, Shandong province conducted a large-scale satisfaction survey in 17 cities on eight topics that people had strong opinions about, such as social security, education, medical care, and health. The results showed that throughout Weifang, education has ranked first in public satisfaction for five years.

"We have been promoting the education reform to this day, and that doesn't mean there are no problems. In the process of deepening the reform, new problems will always be coming up. But we are not afraid anymore, because in Weifang we have formed a comprehensive management system for quality education – 'led by evaluation, with experts running schools, social participation, and a supervision guarantee' – fostering a good environment for the development of education." Guohua Zhang said that as long as we stuck to reform and innovation, stuck to thinking through problems and taking measures concerning students' growth, stuck to problem-solving oriented solutions, stuck to starting from the department's own reforms and straightening out the relationship between politics and the school, and stuck to the path of modern education governance, the road to education reform would get wider and wider, and smoother and smoother.

Ten years of exploration experience have shown that municipal education and manpower is a necessary extension of provincial co-ordination, because the municipal Bureau of Education is in the "meso" level of the national education governance system. Its co-ordination scope is "not wide, not narrow." It cannot only implement the central and the provincial top-level design flexibly and effectively, but also co-ordinate all the resources within its jurisdiction to combine the top-level design with grassroots innovation and promote it in a large area. We can say what the city has achieved is a landmark breakthrough and the only way out for the overall advancement of regional education reform.

It needs to be clarified that "city-based" management does not mean "province-based" or "county-based." The key is to understand what the education administrative departments at different levels are primarily responsible for. Figure this out from role understanding and institutional arrangements, and then the responsibilities of all levels for implementing the development of education will be clear. City-based management can establish an education-integrated linkage mechanism covering a large area, and construct a system for the execution and implementation of a system of education quality promotion. The nature of the solution is to implement creatively, in accordance with national top-level design and provincial deployment requirements.

If the central education decisions and provincial education and manpower are the "brain" and the "arm," respectively, and county education is the "leg," then municipal education is the "spine." If the "spine" is not straight, it will be difficult for the body of national education governance to stand up and walk vigorously. In this sense, the coordinating authority and responsibilities of the Department of Education should be strengthened, to form a connecting link.

With more than ten years of reform, Weifang has built the municipal educational institutions' brand and reputation. "With doing, then with status," their reform has made the local government and the people see that education is no longer the department of "only spending money," and that school is no longer for entrance examinations; instead, they are the important areas in which to create opportunities and value for every student and family and the whole region's economy and society.

## References

[1] Ball, S. J. (2011). *Politics and Policy Making in Education*. [M] Shanghai: East China Normal University Press.

[2] The CPC Central Committee's decision on a number of important issues of deepening reform and opening up in an all-around way [Z]. (2013).

[3] How can Governance System and Governance Capacity Realize Modernization [N]. (2013, December 4). *Guangming Daily*.

[4] Ling, H. (2010). *From public governance paradigm: Research on the functionality changes of local educational administration* [D]. Shanghai: East China Normal University.

[5] Tao, L. (2010). Public Governance of China Education under Globalization–an Elaboration and Criticism of Political Philosophy from the Perspective of Gesellschaft [J]. *Education Research Monthly*, *3*.

# 5

# Analysis of the Current Situation and Strategies for Vocational Education Attraction in Beijing

**Dayong Yuan**

Institute of Vocational and Adult Education (IVAE) at Beijing Academy of Educational Sciences (BAES), Beijing, China

In July 2014, the State Council released "Decision on Accelerating the Development of Modern Vocational Education" to accelerate the development of modern vocational education and further define the important status of vocational education in the national personnel training system. In addition, it asserted that vocational education should meet the requirements of technological progress, production mode reform and social public service, in order to meet the needs of billions of high-quality laborers and technicians. As a development policy, "accelerating the development of modern vocational education" was based on the current disposition of human resources power and entailed speeding up the popularization of high school education, which are wise strategic choices. Beijing always attached great importance to the development of vocational education and put forward a series of policies to promote vocational education. Since the Beijing Vocational Education Conference was held in 2006, financial investment in vocational education has increased significantly, capacity building has been vigorously promoted, and the current vocational education system has been established, following a pattern in which secondary vocational education is the subject and the secondary-to-higher vocational education connection is the focus. Technical and vocational training is widely promoted and public and private vocational education enjoy joint development. Generally speaking, since both central and local governments pay a great deal of attention to vocational education, it should be undergoing substantial development. Yet vocational education recruitment has been relatively weak so far, leading to a state of atrophy.

## 5.1 Current Situation of Vocational Education Attraction

1. The connotations of vocational education attraction

The Modern Chinese Standardized Dictionary defines attraction as the ability to use something's own properties or features to attract the interests or attention of other people or objects. A report co-written by Johanna Lausanne and Jean Gordon, "Enhance Attraction and Social Image of Vocational Education and Training," defines attraction as an attitude of favorable preference and the related behaviors of individuals, groups, and their families toward certain objects. Vocational education attraction and social recognition is shaped by the following factors: increasing the trainees' employment, career development, and promotion opportunities; improving the quality and diversity of learning environments to satisfy the needs of different learners; encouraging citizens and individuals to choose vocational education and training as an approach to self-development; and a willingness to invest in vocational education and training.

2. External determinants of vocational education attraction

External determinants of vocational education attraction include the following: for vocational colleges graduates, to increase employment, career development, and promotion opportunities; for students, to provide high-quality and diverse learning environments in order to satisfy the needs of different learners; for citizens and individuals, a willingness to choose vocational education and training as an approach to self-development; and for enterprises and individuals, a willingness to invest in vocational education and training. Vocational education's attractiveness depends on enrollment and employment. On the one hand, greater selectivity and higher enrollment levels support vocational education; on the other hand, the popularity of vocational education graduates in the job market makes this career path more attractive. Enrollment and employment can be regarded as "import" and "export" issues in vocational education. Exports are perhaps of more importance than imports in this context. Employment opportunities for graduates should be given priority, especially in the current context of economic crisis. Good employment prospects, in turn, contribute to the attractiveness of vocational education, which in turn leads to strong enrollment numbers.

3. Vocational education attraction declines in Beijing

Currently in Beijing (in 2013), there are 122 secondary vocational schools with 218 thousand students; 26 higher vocational schools with 108 thousand students (including vocational education in college and university); and a total of 2.54 million students registered in vocational training institutions. There are 21 private secondary vocational schools and 8 higher vocational schools in Beijing. Vocational education in Beijing has been diversifying, the structure of vocational education has been optimized, and school quality and efficiency have been improved. By 2013, the number of secondary vocational schools had been adjusted from 163 in 2006 to 122, with a rise in the proportion of national and municipal key schools from 51% to 57%. The independent establishment higher vocational schools has increased from 21 in 2007 to 26 today. The number of national demonstration higher vocational schools has reached 27. Vocational school graduates' employment rate exceeds 95%, higher than the national average.

a. Less passionate involvement in vocational education

In recent years, with the booming of popularity of higher education and the development of higher education in Beijing, vocational schools have continued to expand. As one of the main forces in education, vocational education faces several problems, among which one major concern is a lack of attraction. Many students and their parents disapprove of vocational education, which leads to difficulty with recruitment. A student-resources crisis has created a bottleneck in vocational education development. Poor student quality forces schools to lower their teaching standards and start with only basic knowledge. These low teaching standards cannot meet the social requirements of vocational education. In addition, because of national unified enrollment of graduates from vocational schools and general high schools, even higher enrollment levels are required to provide students resources, leading to mixed student quality. This increases the difficulty of teaching and management.

b. Vocational education's inability to attract businesses

Businesses are consumers and beneficiaries of vocational education's products. Clear education targets and strong goal-orientation are what determines the popularity of vocational education in the job market. But a countervailing tendency makes reflection necessary. The key to vocational education development is to make the system employer-oriented, which requires the close involvement of employers and enterprises. Thus far, this type of school-enterprise cooperation needs work, as businesses seldom provide support for job training. Some cooperation arrangements are called "friendship sponsorships," since they are based on personal relationships and lack stable and long-term cooperation mechanisms.

c. The decline of vocational school enrollment in Beijing

Since 2006, the level of enrollment in secondary vocational education has stagnated at around seventy or eighty thousand students (at least 60,800 and at most 83,300). The number of students peaked in 2007 at 262,300. Since then, the number has declined rapidly to 208,600 in 2013. Though enrollment levels in secondary vocational education remains at seventy thousand, household registers of students in vocational education have declined rapidly due to the falling number of middle school graduates and the increasing popularity of high school. Students from outside Beijing contribute half of overall enrollment. Enrollment in higher vocational education has stagnated at forty thousand since 2006, with at most 44,600 and at least 37,700. The number of students peaked at 130,600 in 2007, and then continued dropping to the lowest level in recent years, 108,100 in 2012.

## 5.2  Analysis of the Current Situation of Vocational Education Attraction in Beijing

According to the definition of vocational education given above, vocational education attraction in this study applies to the following groups: (1) people who haven't entered vocational education, including students and parents who may choose vocational education based on certain needs or facts, and (2) enterprises and society, depending on whether the products of vocational education can gain social recognition, attract employers and be welcomed by enterprises.

1. Questionnaire design

Survey methods are one of the most common methods in scientific research, and can be used to assess the present and past situations of related research objects in a systematic, goal-oriented and organized way. Survey methods combine several research approaches, such as historical methods, observational methods, conversations, questionnaires, case studies, and hypothesis tests to achieve a systemic and comprehensive understanding of the research subject. With analysis, comprehension, comparison, and induction using collected data and information, survey methods can provide regularity in the understanding of a given topic. One of the most frequently used methods is a questionnaire, or a one-time use of questions to collect information. Researchers design questionnaire forms based on their research, distribute them to research subjects, and ask them answer the questions. After statistical analysis of the data collected in this way, research conclusions can be drawn.

2. Questionnaire results

In June 2011, the research group conducted an in-depth investigation of vocational education in Beijing, holding several research forums and interviewing many leaders and teachers in vocational education. Based on these in-depth discussions, the research group designed a questionnaire on vocational education attraction in Beijing. They randomly chose 200 vocational school graduates from a vocational school in Beijing as research subjects and asked them to complete the questionnaire. After questionnaire design, pre-testing was conducted using the same group of participants. Generally, the ratio of the number of questionnaire forms used in pre-testing and in actual research should be 1:5. The participants in the study were not graduates. Two hundreds questionnaires were distributed and 187 have been received, with a recovery rate of 93.5%.

In Figure 5.1, the first five items assessed students' overall approval of specific factors related to vocational education attraction: (1) immediate rewards after study; (2) opportunities to learn related skills; (3) opportunities to develop overall abilities; (4) opportunities to gain work experience; and (5) opportunities to gain related professional training. The main reason vocational education attracts students is the practicality of teaching content and the short-term teaching cycle. The first five items assessing students' overall disapproval of factors related to vocational education attraction are: (1) feeling good about oneself in vocational school; (2) teachers are good at communicating with students; (3) innovative activities in school; (4) clear rewards and consequences in school; and (5) the quality of the school's reputation among employers. Vocational education in Beijing has certain general and specific problems, such as feelings about vocational school, communication between teachers and students, and the reputation of vocational schools (Table 5.1).

3. Analysis of external factors that influence vocational education attraction

There are three levels in vocational education attraction: attention, cognition, and action. These levels are correlated, and their influence on attraction increases progressively. The most important measure of vocational education's degree of attraction is the selective behavior conducted by stakeholders. Vocational education attraction can be sub-classified into attraction to vocational academic education and vocational training, respectively. There are different standards with which to measure vocational education attraction, but all of them boil down to whether or not it provides learners with the fundamental

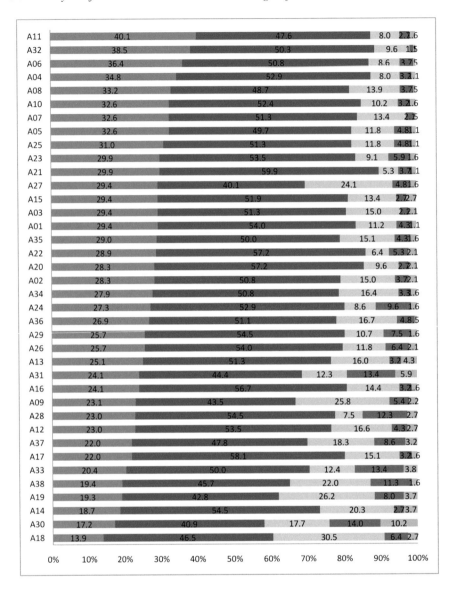

**Figure 5.1**    Students' opinions of vocational education attraction.

**Table 5.1**   Students' opinions of vocational education attraction, in rank order

| Normal Order | Total Approval | Total Disapproval |
|:---:|---|---|
| 1 | To gain immediate reward after study | To feel good about oneself in vocational school |
| 2 | To learn related skills | To enjoy sound communication with teachers |
| 3 | To develop overall ability | To conduct innovative activities at school |
| 4 | To gain working experiences | To be stimulated by the incentive plan at school |
| 5 | To receive related professional training | To be more competitive in employment thanks to the reputation of the training school |

tools they need to develop in a vocational career. The stakeholders include: the public, who may or may not choose vocational education, but pay attention to it, developing a common public opinion; the government, which determines the status of vocational education, the distribution of resources, institutions, and policies, leads vocational education development and shapes attraction; enterprises and corporations, whose involvement (and the depth and breadth of this involvement) decides the quality and features of vocational education, and whose acceptance of vocational school graduates directly determines vocational education attraction; the schools and institutions of vocational education, whose teaching quality determines attraction; and students and parents, whose choices ultimately determine the survival and development of vocational education.

a. Social factors that influence vocational education attraction

Compared to general education, vocational education is less attractive because of low expectations about income and social status, which will determine future social stratification and occupational stratification. Occupational stratification is an important basis of social stratification. According to M. Weber's theory, there are three criteria with which to classify professional classes: income (economic income), status (honor and reputation), and power (ability to control resources). Within vocational education, the attractiveness of professional education in different occupations varies. Vocational education offers training in a variety of occupations, with in different vocational schools corresponding to different occupations; their majors, enrollment, and employment levels vary greatly. Vocational education is by nature a type of job training. National and local labor, employment, personnel systems, and policies have an indirect but profound impact on vocational education attraction. On the one hand, they can guarantee good, stable employment for vocational school graduates; and this can stimulate a virtuous cycle, in which education outcomes demonstrate the effectiveness and value of vocational education, contributing to attraction. The new government in Beijing has adopted a new orientation for this city: Beijing will be developed into the center of politics, culture, international communication, and scientific innovation. With great population pressure and this new orientation, irrelevant functions and organizations will be squeezed out. Industrial reforms and transformation have introduced more variables to the fragile field of vocational education in Beijing.

### b. Educational factors that influence vocational education attraction

National education policy and development plans, especially vocational education structural adjustment policies in high school education and funding policies, have had a decisive influence on the expansion of demand for vocational education. These policies also represent an important opportunity to make vocational education more attractive. Due to the impact of the economic situation and the university enrollment expansion policy and its quality, vocational education has undergone five years of negative growth since 1998 and its attraction has begun to decline. The state council released a "Decision on Promoting the Reform and Development of Vocational Education" in 2002, proposing that the scale of secondary vocational education should be comparable to that of general high school education. This policy has reversed the decline of vocational education. But due to its own incomplete development, vocational education can be characterized as "beheaded education", in that students in vocational education are confined to a lower level of training and lack opportunities for social mobility. Currently, vocational education hasn't developed a practical frame of reference or institutional design compatible with other levels of education, and cannot open up further opportunities for career development. The status and reputation of vocational education is shaped by the education system itself, namely administrative departments and schools. In reality, the high school graduation rate is the more important indicator of education quality.

### c. Factors within vocational education itself

For students and parents, there are at most seven factors that impact the choice of vocational education: reputation, major structure, education cost, teaching quality, employment, entrance opportunities in higher education, and incentive policies. Whether the major structure matches the local economy is an influential factor, which depends on whether vocational education can satisfy the needs of the local industrial structure. Major structure and talent development should be based on national industrial development and local economic development to make sure that the major matches the industry, the major chain matches the industrial chain, and the school matches the local economy. Vocational education graduates can therefore serve local development and be welcomed by local enterprises, which in return guarantee their employment. Entrance exam systems and well-established systems also impact vocational education attraction. At present, the proportion of students entering higher education is controlled by national policy, which prohibits certain students from entering higher education before finding a job. In addition, due to the different orientation of secondary and higher vocational education, there are systemic obstacles to linking the two systems. Breaking through these systemic obstacles, adjusting entrance policies and entrance proportions in different regions, perfecting the vocational education system, and linking secondary and higher vocational education can encourage and guarantee students' choice of vocational education at an appropriate age. To implement testing in a national education reform pilot project, Beijing introduced a grading system reform test in vocational education in 2011 and a "three plus two" link between secondary and higher vocational education in order to explore the modern vocational education system. Reforming training practices, with approaches like the new "technique plus basis" model of adult secondary academic education in rural areas, and a "credit bank," a flexible education program designed for enterprise workers, has improved vocational education attraction. Meanwhile, teaching quality, features, funding, and incentive policies have become important factors in vocational education itself, encouraging urban and rural students from poor families to choose vocational education. Statistics show that 80% of vocational education students come from countryside and 20% to 30% of urban students come from low-income families. There are clearly economical obstacles for poor students entering vocational education due to cost sharing.

## 5.3  Policy and Advice to Improve Vocational Education Attraction in Beijing

1. In education content: develop rich, in-depth content in vocational education

To improve the quality of vocational education, "education value" should be stressed. In job training, emphasis should be placed on the training target of vocational ability and improving basic literacy, professional competence, and professional knowledge. This entails distinguishing and properly orienting training targets at various levels of vocational education, and establishing a dynamic mechanism for selecting majors based on the requirements of the labor market, employers, and professions throughout the country. Following a full discussion and investigation of the major structure in vocational education, scientific specialty catalogues and guidelines should be released. Taking into consideration local and school-specific situations, this process should determine clear major orientations and school features. Education administrative departments should offer effective guidance in setting the majors for vocational education, to coordinate with regional professional arrangements. Innovative education platforms should be integrated into the vocational education system in order to develop a lifelong vocational education system. The government should transfer part of its vocational academic education funding to vocational training in order to adapt to the improving status of vocational training. There should be various types of vocational training that conform to the government's guidance and social needs, and for those offering training to disadvantaged groups, such as laid-off and migrant workers, the government should provide funding.

2. In education management: improve environmental support and build bridges for talent growth

It is also important to promote institutional reform and optimize the development environment of vocational education in Beijing. The success of secondary vocational education will lead to the integration and coordinated development of secondary and higher vocational education, which comprise vocational academic education in Beijing. Vocational education should develop a holistic approach to promoting coordination among the different layers of vocational education. Enhancing the link between vocational education and general education and handling its relationship with other education institutes are inevitable requirements for its development. Beijing has entered into a stage of overall education reform. In its relationship with general education, adult education and higher education, vocational education should develop innovatively, encourage industry and enterprise, and participate in and conduct vocational education. The key to the innovative development of institution design in vocational education is promoting enterprise involvement in vocational education and forming long-term mechanisms. Flexibility, openness and richness are basic features and requirements of vocational education. Lifelong vocational education and vocational education for all is the goal for its development. This means breaking the glass ceiling and building more links to industry and the labor market, and combining and integrating vocational education into general education and the university system.

3. In education subject: transforming government functions, conducting classified management, and exploring various modes of development tailored to business needs

The government should transform its functions and improve its management structure, which requires the government to handle its relationship with schools, training organizations, enterprises, and the public, as well as its relationship with the market. The government should be transformed from a provider of vocational education into a purchaser of its services. It should also be in charge of making related policies and regulations, guiding and propagating vocational education, strengthening planning and supervision, adjusting and offering information support, etc. Education management departments should offer classified management to vocational education, which can be sub-classified into vocational academic education and non-academic education; vocational school education and vocational training; and pre-service education, post-service education, and in-service education. All of these categories comprise vocational education. The subjects of vocational education are not confined to vocational schools, social training organizations, industries, and enterprises; intermediary organizations and even families can conduct vocational education. Their relationships should be handled properly, arranging functional divisions in vocational education. With the most direct and closest link being that between vocational education and the economy, between enterprise and market, the inevitable path for development is market-orientated and enterprise-based, a combination of business and education. Intermediary organizations such as *guilds* and *boards of directors* should be established to transform these factors from external relationships into actual organic interactions.

## References

[1] Fletcher, M. (2012). Effective transitions from school to work: the key role of FE colleges. London: 157 Group.
[2] Attractiveness of initial vocational education and training: identifying what matters/European Centre for the Development of Vocational Training [Ed.]. – Luxembourg: Publications Office of the European Union, 2014.

[3] Lovsin, M. (2014). The (un)attractiveness of vocational and techonical education: theoretical background. Ljubljana, Slovenia: Center for Educational Policy Studies, 2014. *CEPS Journal* 4:1, pp. 101–120.

[4] Ratnata, W. (2013). Enhancing the image and attractiveness of TVET. *VET@Asia* 1, pp. 1–13.

[5] Smith, E., Kemmis, R. B. (2013). Good practice principles in apprenticeship systems: An international study. *TVET Asia* 1, pp. 1–12.

[6] Zuguang, Y. (2009). Improve the consistency of the development of vocational education and social needs [J]. *Chinese Vocational and Technical Education* 18, pp. 6–8.

[7] Shuchao, M. (2009). Vocational education must break through the bottleneck of development must enhance the attractiveness [J]. *Vocational and Technical Education* 30:09, p. 22.

[8] Xin, J. (2009). Efforts to complete the task of vocational education reform and innovation to promote quality improvement [N]. *China Education News*, 2009-08-03.

[9] Ministry of Education. (2007–2012). China National Education Development Statistics Bulletin.

# 6

# Academic Institutions and Their Operation Modes in University Charters

— Textual Analysis of 15 University Charters Reviewed and Approved by the Ministry of Education

**Yu Hong**

Hebei University, Beijing, China

## Abstract

The Ministry of Education and the provincial education administrative authorities will finish the review and approval of all university charters by the end of 2014, so the preparation, review and approval of university charters is currently an urgent task. University charter preparation entails a multitude of tasks addressing many aspects and problems with education, and one of the most important problems is how to establish a governance system to which modern academics can adapt. Through a textual analysis of 15 university charters approved by the Ministry of Education in June, it can be seen that the university charters have several issues, including changes in their legal basis, vague expressions, empty statutory rights, unclear relationships, and loose flow or poor convergence in terms of regulations of academic institutions and their operating modes. The urgency of these issues is obvious, and the education administrative authorities and universities should establish links with each other and work closely to address them. The universities, which prepare the charters, should adhere to academic autonomy, pursue dual-legitimacy, and prepare rigorous and practical provisions; and the education authorities at all levels should provide macro-level direction, set appropriate approval standards, and leave enough room for innovation.

**Keywords:** University charter, academic governance system, academic autonomy, academic governance by professors.

## 6.1 Introduction

The General Office of the Chinese Ministry of Education issued a "Notice on Accelerating the Development, Approval and Implementation of University Charters" on May 28, 2014, requiring the universities in "985 Project" and "211 Project" to submit draft university charters to the Ministry of Education before June 15 and November 30, respectively. The Ministry of Education and the provincial education administrative authorities will finish the review and approval of all university charters for these universities in "985 Project" and "211 Project" by the end of 2014, and for all universities by the end of 2015. Therefore the preparation, review and approval of university charters is an urgent task at present.

To review the evolution of the policies related to university charters, they comprise the "constitutions" of universities and also the foundations of academic autonomy. The establishment of university charters has been a top priority in the improvement of the legal system of higher education and the construction of a modern university system in China. The "Education Law of the People's Republic of China" (hereafter referred to as the "Education Law") was promulgated in 1995, and the "Higher Education Law of the People's Republic of China" (hereafter referred to as the "Higher Education Law") was promulgated in 1998. These laws clearly specify that schools shall teach autonomously according to their charters; and the "Opinions of the Ministry of Education on Strengthening the Legal System of Education" further requires universities to "develop and improve their charters as soon as possible according to the provisions of laws and regulations." Even so, only a very small number of public universities have their own charters as of 2010, and the consequences include vague legal issues, unclear authority boundaries between the sponsors and universities, and unclear rights and obligations between teachers and students and the universities (Xiangju, 2004). It was not until 2012 that the Ministry of Education promulgated the "Provisional Measures on the Preparation of University Charters", which has clarified the contents of the university charters, preparation procedures, review and approval, supervision, and other key issues. Since then, university charters have been prepared according to these regulations and are on the right track. At present, the charters of 15 universities have been reviewed and approved by the Ministry of Education (Xiangju & Guangli, 2004).

University charters cover many issues, primarily the transformation of government functions, the improvement of external social supervision, and the optimization of governance structures and the strengthening of academic power within the universities. Among these issues, problems related to academic power have been a difficult point in policy practice and a hot topic of academic research. On the one hand, excessive interference by the administration is a chronic problem plaguing the development of universities in China, and the crux of the issue lies in the aggressive use of administrative power, with universities being "arranged" by administrative departments, thereby marginalizing the power of academics and limiting the abilities of academic institutions. It is therefore urgent that universities take this opportunity to use their charters to alter the disposition of academic power in the university, giving primacy to professors involved in teaching and research activities. On the other hand, the roles and functions that academic power plays in university governance are controversial among theorists, and there is divergence between the two concepts of "university administration by professors" and "academic governance by professors" (Yanghong, 2012). The expression "academic governance by professors" is widely adopted in the university charters that have been approved by the Ministry of Education thus far. This has caused some problems that should be studied further.

A text analysis of 15 university charters, which have already been reviewed and approved by the Ministry of Education, has been carried out, and the legal basis, functions, rights and responsibilities of the academic institutions specified in the charters are discussed in-depth in this paper. Together with case analyses of other universities, the existing problems in the academic management systems of universities in China are discussed and summarized, with suggestions on how to resolve these problems. On the one hand, this paper is a response to key theoretical issues involved in the establishment of the academic system in modern universities; on the other, it can help the universities circumvent errors in the preparation of their charters, and enhance the rigor and scientific nature of the system's design.

## 6.2 Academic Institutions and Their Functions in 15 University Charters

An analysis of the academic institutions specified in 15 university charters shows that all universities have Academic Committees; Academic Degree Evaluation Committees are clearly specified in 14 university charters, Teaching Committees are available in 12 university charters,[1] Title and Job Duty

---

[1] The names of "Teaching Committees" vary in different university charters. For example, the Renmin University of China calls this the "Talent Training Committee"; Shanghai International Studies University, Sichuan University, and Shanghai University of Finance and Economics call it the "Teaching Guidance Committee"; Tongji University and Northeast Normal University call

Evaluation Committees are available in 7 university charters,[2] and Academic Ethics Committees are available in 3 university charters.[3] In addition, there are some rare academic institutions, such as the Discipline Establishment Committee in the charter of Shanghai University of Finance and Economics, and the Academic Evaluation Committee in the charter of China University of Mining & Technology (see the following Table 6.1).

1. Academic Committee

The preparation of university charters is guided by existing laws. As a standing body in all universities, the establishment of the Academic Committee is guided by four legal provisions: first, the "Higher Education Law" promulgated in 1998, of which Article 42 states: "Universities shall establish Academic Committees to review the establishments of disciplines and majors, review teaching and scientific research programs and plans, and assess the teaching and scientific research results"; second, the "Under-graduate Major Establishment Regulations" promulgated in 1999, of which Article 24 states: "the Academic Committees ... review the major establishment and adjustment in the universities"; third, the "Methodology of Preparation of University Charters" promulgated in 2012, of which Article 11 states: "The charters shall clearly specify the principles of organization, selection of persons responsible, operating rules and supervision mechanisms of the Academic Committees and Academic Degree Evaluation Committees in universities, to ensure that the academic institutions ... play their roles in consultation, review and decision making; and fourth, the "Regulations on

---

it the "Senate Committee"; and Southeast University, Donghua University, Wuhan University of Technology, Jilin University, Shanghai Jiao Tong University, and China University of Mining & Technology call it the "Teaching Committee." Since the functions of the committees above are identical, all are referred to as "Teaching Committees" here.

[2]The names of "Title and Job Duty Evaluation Committees" also vary in different university charters. Southeast University calls this the "Title Evaluation Committee," Wuhan University of Technology calls it the "Senior Professional Skill and Job Duty Evaluation Committee," Jilin University calls it the "Teacher Job Duty Evaluation Committee," Shanghai Jiao Tong University calls it the "Professional Skill and Job Duty Evaluation Committee," Tongji University calls it the "Technical Skill Employment Committee," and Sichuan University and Shanghai University of Finance and Economics call it the "Professional Skill and Job Duty Evaluation Committee." Since the functions of the committees above are identical, they are all referred to as "Title and Job Duty Evaluation Committees" here.

[3]Shanghai Jiao Tong University has established an "Academic Discipline and Ethics Committee," which is very similar to the "Academic Ethics Committee" established in Shanghai University of Finance and Economics and China University of Mining & Technology, therefore these are all referred to as "Academic Ethics Committees" here.

**Table 6.1** Academic institutions and their operation modes in university charters

| University | Academic Committee | As the Highest Academic Institution or Not | Academic Degree Evaluation Committee | Teaching Committee | Title and Job Duty Evaluation Committee | Academic Ethics Committee | Discipline Establishment Committee | Academic Evaluation Committee |
|---|---|---|---|---|---|---|---|---|
| Renmin University of China | ✓ | No | ✓ | ✓ | × | × | × | × |
| Southeast University | ✓ | No | ✓ | ✓ | ✓ | × | × | × |
| Donghua University | ✓ | No | ✓ | ✓ | × | × | × | × |
| Shanghai International Studies University | ✓ | Yes | ✓ | ✓ | × | × | × | × |
| Wuhan University of Technology | ✓ | Yes | ✓ | ✓ | ✓ | × | × | × |
| Central China Normal University | ✓ | Yes | ? | ? | × | ? | ? | × |
| Jilin University | ✓ | Yes | ✓ | ✓ | ✓ | × | × | × |
| Shanghai Jiao Tong University | ✓ | Yes | ✓ | ✓ | ✓ | ✓ | × | × |
| Tongji University | ✓ | Yes | ✓ | ✓ | ✓ | × | × | × |
| Sichuan University | ✓ | Yes | ✓ | ✓ | ✓ | × | × | × |
| Northwest A&F University | ✓ | Yes | ✓ | ? | × | ? | ? | × |

(*Continued*)

**Table 6.1** Continued

| University | Academic Committee | As the Highest Academic Institution or Not | Academic Degree Evaluation Committee | Teaching Committee | Title and Job Duty Evaluation Committee | Academic Ethics Committee | Discipline Establishment Committee | Academic Evaluation Committee |
|---|---|---|---|---|---|---|---|---|
| Northeast Normal University | √ | Yes | √ | √ | × | × | × | × |
| Shanghai University of Finance and Economics | √ | Yes | √ | √ | √ | √ | √ | × |
| China University of Mining & Technology | √ | Yes | √ | √ | × | √ | × | √ |
| Southwest University | √ | Yes | √ | × | × | × | × | × |

*Note:* These universities, from the Renmin University of China at the top to the Southwest University at the bottom, are listed in order of approval serial numbers assigned by the Ministry of Education. The 9 universities below the double line are the universities whose charters were approved by the Ministry of Education in the 2nd round. "√" means that the institutions are available in the university charters; "×" means that the institutions are not available in the university charters; and "?" means that the university charters specify establishing "a number of specific committees" for handling related matters, but don't provide the names.

Academic Committees in Universities" (hereafter referred to as the "Regulations on Academic Committees") promulgated in 2014, of which Article II states: "Universities should ... improve the academic management systems and organizational structures with Academic Committees at the core, and should regard the Academic Committee as the highest academic institution in the universities, generally exercising the rights including decision making, review, evaluation, and advising related to academic affairs within the universities. These regulations also specify that Academic Committees are entitled to review and determine the following affairs: key academic planning, appointment of teachers with different titles, academic evaluations, academic controversy resolutions, and the preparation and modification of charters of specific committees and sub-Academic Committees. In addition, the Academic Committees are entitled to approve the following affairs: awards for teaching and research achievements, the introduction of high-level talent, the selection of personnel in important academic institutions, and the evaluation of independent academic projects and awards; and they are entitled to provide advice on the following matters: strategic planning related to the universities and academic affairs, allocation and utilization of teaching and research funds specified in the budgets, application of key projects and allocation and utilization of funds, education projects founded between Chinese universities and foreign universities, and important foreign cooperation projects. It can be seen through a comparison of these laws and regulations that the "Higher Education Law" promulgated in 1998 did not designate the Academic Committees as the highest academic institutions; at that time, their functions were limited to review and discussion of relevant academic affairs. The "Methodology of Preparation of University Charters" promulgated in 2011 and the "Regulations on Academic Committees" promulgated in 2014, however, have endowed the Academic Committees the "rights of decision making" critical to some academic affairs, and the "Regulations on Academic Committees" further clarifies that the Academic Committees are the "highest academic institutions" at the core of academic institutions. This change reflects how policymakers and legislative promoters, headed by the Ministry of Education, have gradually gained a deeper understanding of "academic governance by professors" and reached some consensus in this regard: with the Academic Committee at the core, the expansion of empowerment and the centralization of power in the academic institution can guarantee that lead roles are played by the scholars in decision making on academic affairs, and thus internal checks and balances on the university administration can be realized.

The Academic Committees are appointed as the highest academic institutions in most of the charters. In addition to the charters of the Renmin University of China, Southeast University, Donghua University, Central China Normal University, Sichuan University, Northwest Agriculture and Forestry University, Northeast Normal University, Shanghai University of Finance and Economics, China University of Mining and Technology, and Southwest University that specify the Academic Committees as the "highest academic institutions," the other five universities also set the Academic Committees as the "highest academic authorities" or "highest decision-making bodies on academic affairs" in their charters. In the charters of three universities, including the Renmin University of China, the Academic Committees aren't affiliated with other academic institutions within the universities, but are instead considered parallel institutions. The reason why the charters of these three universities are different from those of other universities may lie in their earlier submissions of their charters to the Ministry of Education (reviewed and approved in November 2013), when they could not refer to the latest "Regulations on Academic Committees" (adopted in January 2014). These charters are instead primarily based on the Higher Education Law, limiting the functions of the Academic Committees to four areas, including the review of the establishments of academic institutions, the review of the establishments of disciplines and majors, the review of academic development planning, and the evaluation of teaching and scientific research results. Some universities directly copy clauses in the "Regulations on Academic Committees" in their charters. Central China Normal University, Northwest Agriculture and Forestry University, and Shanghai University of Finance and Economics specify duties for Academic Committees that are highly consistent with the "Regulations on Academic Committees" promulgated by the Ministry of Education. Of these, Central China Normal University was informed about part of the "Regulations on Academic Committees" prior to November 2013, and basically copied the part about the duties of the Academic Committees into its charter; this was one of the earliest charters to be approved, and the style of this section, showing obvious signs of modifications, is very different from that of the other parts of the charter. The decision-making rights of Academic Committees are not clearly specified in most of the university charters. In the six university charters reviewed and approved in the first round, except for the charter of the Central China Normal University, no decision-making rights of Academic Committees are specified; while in the 9 university charters reviewed and approved in the second round, except for the charter of the Tongji University, the statements in the "Regulations on Academic

Committees" have been adopted in the university charters, specifying that the Academic Committees are entitled to carry out "decision making, review, evaluation, and consultation" for academic affairs. In summary, including the Central China Normal University, it is clearly specified in a total of 9 university charters that the Academic Committees enjoy rights of decision making, but through careful reading of the specific clauses in these 9 university charters, it can be seen that specific decision-making rights are defined in only 3 university charters: the Academic Committee of Sichuan University is entitled to "formulate the policy documents of the Research Fund and make decisions on funding field and direction, as well as the appointing important projects"[4]; the Academic Committee of Northeast Normal University is entitled to "review and approve academic evaluation criteria"[5]; and the Academic Committee of Southwest University is entitled to review and approve teaching and scientific research achievements, talent training quality evaluation criteria, academic degree-granting criteria, education programs, employment criteria of teachers with professional and technical titles, the charters of specific committees and sub-Academic Committees, academic evaluation criteria, and academic controversy resolutions. Some other universities only copy the expressions specified in the "Regulations on Academic Committees"; although they specify the affairs involved in "review and decision making" by the Academic Committees, they fail to specify the affairs that can be "reviewed" only, as opposed to the affairs can be directly "decided," so in these charters, vague expressions will make it difficult for the Academic Committees to exercise their decision-making rights. Some universities specify that the Academic Committees enjoy the decision-making rights in their charters, but include no clauses related to these rights; for example, in the 10 clauses related to the duties and responsibilities of the Academic Committee in the charter of the China University of Mining and Technology,[6] the first 5 clauses and the 7th clause are related to "review," the 6th and 8th clause specify rights of "guidance and supervision" on relevant bodies, the 9th clause specifies the rights of "evaluation" in academic controversies, and the 10th clause specifies duties to be performed and other tasks assigned. None of these 10 clauses assigns direct decision-making rights to the Academic Committee.

Concerning the membership of Academic Committees and their qualifications, the charters of Southeast University, Wuhan University of Technology,

---

[4]Clause 3, Section 47 in "Charter of Sichuan University".

[5]Clause 3, Section 26 of "Charter of Northeast Normal University".

[6]Clause 53 of "Charter of China University of Mining and Technology".

Tongji University, and Sichuan University clearly specify that the Academic Committee members should be experts and scholars with high academic prestige. The charters of Northwest University of Agriculture and Forestry, Shanghai University of Finance and Economics, China University of Mining and Technology, and Southwest University require that the Academic Committee members should be senior professors. The charter of Shanghai Jiao Tong University is different from the others, instead specifying that Academic Committee members should be divided into two categories: functional members and elected members, of which functional members should be relevant university leaders, and the elected members should be well-known professors elected democratically. In terms of the appointment of members, according to the seats corresponding to the colleges (disciplines), the Academic Committee members are appointed based on bottom-up democratic elections (matching the settings of disciplines and majors in the universities) in the Academic Committees of Jilin University, Shanghai University of Finance and Economics, and Southwest University. The Academic Committee members are appointed in elections by sub-Academic Committees in Wuhan University of Technology: candidates for Academic Committee membership are nominated by grassroots academic organizations, discussed and determined in a President's Office Meeting (or University Meeting), and then approved by the president (or by the CPC Standing Committee of the university) in Northwest Agriculture and Forestry University and China University of Mining and Technology. Finally, in the Academic Committee of Shanghai Jiao Tong University, functional members are simply appointed and changed according to administrative duties, while elected members, as in the case of Jilin University, are appointed in bottom-up democratic elections.

Concerning the appointment of a committee chairman, the committee chairman is nominated by the president and elected by all members at Jilin University, Shanghai Jiao Tong University, Northwest Agriculture and Forestry University, and Shanghai University of Finance and Economics, and the appointment of the committee chairman can be based on one of two approaches. The first is to be "nominated by the president and elected by the members," and the second is via direct election by all members, as in Southwest University. The committee chairman can also be appointed based on election by members in Wuhan University of Technology. In addition, the committee chairman shall be a senior professor who is not an administrator in the university as specified in relevant provisions in the Renmin University of China; and the committee chairman shall be its one member selected in Southeast University.

Concerning the proportion of administrative staff, Shanghai Jiao Tong University, China University of Mining and Technology, and Southwest University directly refer to the provisions in the "Regulations on Academic Committees" promulgated by the Ministry of Education, specifying that the number of members playing the functions of the CPC party and administration in the universities shall not exceed a quarter of the total number of members, and professional teachers not playing the functions of the CPC party and administration, or main persons in charge of colleges and schools, shall make up at least 1/2 the total number of members, while a certain proportion will be young teachers.

### 2. Ademic Degree Evaluation Committee

The Academic Degree Evaluation Committee is established on the basis of two legal provisions: first, the "People's Republic of China Regulations on Academic Degrees" (hereafter referred to as the "Regulations on Academic Degrees") and the "People's Republic of China Provisional Implementation Measures of the Regulations on Academic Degrees" (hereafter referred to as the "Implementation Measures of the Regulations on Academic Degrees") promulgated in 1980. Of these, the relevant provisions are Article 9 of the "Regulations on Academic Degrees," which specifies: "Academic degree-granting institutions shall establish Academic Degree Evaluation Committees and organize thesis oral examination committees for relevant disciplines," and Article 11 in "Methodology of Preparation of University Charters" promulgated in 2011 (see above). According to the "Regulations on Academic Degrees" and the "Implementation Measures of the Regulations on Academic Degrees," the Academic Degree Evaluation Committee has four rights: 1) to review the lists of bachelor's, master's, doctoral, and honorary doctoral degrees conferred; 2) the determination of the subjects, number, and scope of master's and doctoral examinations, and the review and approval of the lists of key examiners and members of the thesis oral examination committees; 3) determination of granting master's degrees and doctoral degrees; and 4) the resolution of controversies related to academic degrees, including the revocation of the degrees granted if found in violation of relevant provisions. From a chronological point of view, the Academic Degree Evaluation Committee (1980) was proposed earlier than the Academic Committee (1998) in relevant legal documents, and we can say that the Academic Degree Evaluation Committee is the earliest academic institution established in universities since the reopening of the College Entrance Examination. The status of the Academic Committee was as same as that of the Academic Degree Evaluation Committee in the "Methodology of Preparation of University Charters" promulgated in 2011, and it was not the "highest academic institution" until the recent promulgation of the "Regulations on Academic Committees." But in management practice, the Academic Committee is not necessarily stronger than the Academic Degree Evaluation Committee, which was established as the "earliest academic institution" in terms of influence over academic affairs in the universities.

Concerning the rights and responsibilities undertaken by the Academic Degree Evaluation Committee, in addition to the four kinds of rights in the "Regulations on Academic Degree", additional rights and responsibilities are assigned to the Academic Degree Evaluation Committee in some university charters: 1) adjustment of disciplines and majors, in the charters of the Renmin University of China, Wuhan University of Technology, Tongji University, and Northeast Normal University, which specify that the Academic Degree Evaluation Committee is entitled to "review the setting and adjustment of disciplines and majors in universities"; 2) development of discipline planning – the Academic Degree Evaluation Committees of Sichuan University and Northwest Agriculture and Forestry University are entitled to "develop university discipline planning" or "develop academic degree and postgraduate education development plans and related policies"; 3) selection of postgraduate tutors – the Academic Degree Evaluation Committees of the Renmin University of China, Shanghai International Studies University, Wuhan University of Technology, Sichuan University, Northwest Agriculture and Forestry University, and Northeast Normal University are entitled to select the postgraduate tutors or develop selection standards for postgraduate tutors. Concerning the appointment of the committee chairman, the charters of Wuhan University of Technology and Southwest University specify that the chairman should also be the president of the university, and the charter of Shanghai International Studies University specifies that the chairman can be a key person and a senior professor in the university. In practice, presidents are also the committee chairmen in most of universities; the presidents can award academic degrees to the graduates and are responsible for education quality.

Concerning the relationship between the Academic Degree Evaluation Committee and Academic Committee, the Academic Committee is not the highest academic institution in the charters of the Renmin University of China, Southeast University, and Donghua University, so the Academic Degree Evaluation Committees are parallel to the Academic Committees in these three universities. In the other 12 university charters, although the Academic Committee is defined as the highest academic institution, the relationship between the Academic Degree Evaluation Committee and the Academic

Committee is quite subtle; some universities place the Academic Degree Evaluation Committee under the jurisdiction of the Academic Committee. For example, the charter of China University of Mining and Technology specifies that the Academic Degree Evaluation Committee is a specific committee under its Academic Committee; the charter of Central China Normal University specifies that the Academic Committee is entitled to establish and empower a specific committee to deal with the affairs related to academic degree evaluation; and the charters of Shanghai International Studies University, Wuhan University, and Tongji University specify that the Academic Committee is entitled to nominate the members in the Academic Degree Evaluation Committee, and to review the matters reported by the Academic Degree Evaluation Committee. But a small number of university charters fail to clarify the relationship between the Academic Degree Evaluation Committee and the Academic Committee. For example, the charters of Shanghai University of Finance and Economics, Jilin University, and Southwest University specify that the Academic Committee is the highest academic institution, having the rights to review degree-granting criteria and evaluation rules, but the Academic Degree Evaluation Committee is parallel to the Academic Committee in specific clauses – for instance, the charter of the Shanghai University of Finance and Economics defines the academic institution as "including the independent teaching and research institutions, Academic Committee and its specific sub-committees, and Academic Degree Evaluation Committee established by the university," of which the Academic Degree Evaluation Committee is obviously not a "specific sub-committee" of the Academic Committee. Article 35 of the charter of Jilin University specifies that "the university supports the Academic Committee and the Academic Degree Evaluation Committee to respectively exercise their duties and rights according to their own charters," and Article 32 of Southwest University also specifies that "the Academic Committee and the Academic Degree Evaluation Committee is established in the university" and describes each institution's duties, rights, and procedural rules in two sections respectively. The status of the two committees is relatively equal in these three universities. In addition, in the university charters not yet reviewed and approved by the Ministry of Education,[7] the charter of Beijing Normal University specifies that the Academic Committee is the "highest academic institution in the university" while it defines the Academic Degree Evaluation Committee as the "highest authority for academic degree evaluation, granting, and revocation as well as

---

[7]Source: "Compilation '985 Project' University Charters".

controversy resolution," and the chairman is the president. In matters related to academic degrees, the Academic Committee should have less power than the Academic Degree Evaluation Committee, and thus the so-called "highest academic institution" may become a nominal title.

### 3. Teaching Committee

As a specific committee established under the Academic Committee, the Teaching Committee (or Talent Training Committee/Teaching Guidance Committee) has been established in 12 out of the 15 universities whose charters have been reviewed and approved. Obviously, the Teaching Committee plays a critical role in assisting the university in fulfilling its mission of training talent. Therefore, the establishment of Teaching Committees is a general practice in most of the universities.

The legal basis of the establishment of the Teaching Committee is Article XI in the "Regulations on Academic Committees" promulgated by the Ministry of Education: "The Academic Committees can set up a number of specific committees for discipline establishment, teacher appointment, teaching guidance, scientific research and academic ethics." Moreover, Teaching Committees are also established to correspond to the Higher Education Teaching Guidance Committee established in the Ministry of Education. As specified in the "Charter of Higher Education Teaching Guidance Committee of the Ministry of Education," issued in 2006, the Higher Education Teaching Guidance Committee of the Ministry of Education is "an expert organization under the leadership of the Ministry of Education, carrying out research and evaluations and providing advice, guidance and services for teaching in universities." Its functions include: research on the macro-level key issues in higher education reform and development and providing advice to the Ministry of Education and universities; formulating professional norms, teaching quality criteria, and basic teaching requirements for foundation courses and the basic conditions of experimental courses; research on the structure and layout of majors, review of the setting of majors in universities and verification of the setting of majors in colleges; review and promotion of relevant teaching reform programs and achievements; guiding the establishment of disciplines and majors, courses, textbooks, training approaches, and laboratories; supervising and evaluating the teaching quality of relevant disciplines; and organizing teacher training, to exchange experiences on teaching and reform.

For a long time, the education management system has been a highly centralized hierarchy in China, and the settings of the functional departments within universities usually correspond to the functional departments within the Ministry of Education in order to facilitate communication. For this reason, most universities have established their own Teaching Committees corresponding to the Higher Education Teaching Guidance Committee of the Ministry of Education. One problem with this has been presented to us: can this "identical organization" to that of the Ministry of Education be helpful to the modern university's development? If too much attention is paid to following the Ministry of Education in the management of universities, while ignoring their own circumstances and actual situations, this may produce a lack of initiative and autonomy in systematic establishments and organizational structures, and internal management mechanisms may be ossified and bureaucracy may prevail, which would in turn lead to difficulty adapting to the ever-changing external environment, making it impossible to reach the ultimate goal of improving education quality and educational level.

Among the charters of most universities, the duties, rights and responsibilities of the Teaching Committees can be classified into the following four categories: review of the planning for key projects and programs in talent training and teaching reform; review of the regulations and systems related to education management, quality monitoring, and supervision of implementation; review of evaluation criteria and methodology for teaching awards; and ruling on teaching accidents and teaching appraisal controversies. In addition to these four functions, a small number of universities assign additional duties and rights to the Teaching Committees, which include: 1) review of the allocation and utilization of teaching funds – the Teaching Committees in Shanghai International Studies University and Wuhan University of Technology are entitled to review the budget and utilization of teaching funds; 2) verification of the setting and adjustment of majors and disciplines – the Teaching Committee in Shanghai International Studies University is entitled to study and verify the setting and adjustment of majors and disciplines; and 3) evaluation of academic titles – the Teaching Committee in Northeast Normal University is entitled to put forward the job duties and evaluation methods corresponding to academic titles. Concerning the qualifications of the members of the Teaching Committees, clear provisions are provided in only four university charters: the Teaching Committee in Shanghai International Studies University is composed of senior professional and technical personnel who are good at teaching and have the abilities to perform their duties; the Teaching Committee in Tongji University is composed of teachers and some management bodies, students, and employers; the Teaching Committee in Shanghai University of Finance and Economics is composed of teachers who have major academic achievements and bring rich experience in teaching and come from the teaching departments and relevant functional departments in the university; and there is a Supervision Board, composed of retired teachers, under the Teaching Committee in Northeast Normal University.

### 4. Title and Job Duty Evaluation Committee

The Title and Job Duty Evaluation Committee (also known as the Technical Skill Employment Committee, Professional Skill and Job Duty Evaluation Committee, Senior Professional Skill and Job Duty Evaluation Committee, etc.) is the specific committee responsible for title and job duty evaluation for teachers and professional technicians. This kind of committee is clearly specified in a total of 7 university charters.

The legal basis of the Title and Job Duty Evaluation Committee is Article 11 in the "Regulations on Academic Committees," the "Provisional Regulations on Job Duties of Teachers in Universities" (hereafter referred to as "Provisional Regulations on Job Duties of Teachers"), and the "Charters of Job Duty Evaluation Institutions for Teachers in Universities" (hereafter referred to as "Charters of Job Duty Evaluation Institutions for Teachers") issued by the Ministry of Education in 1986. Of these, Article 14 in "Provisional Regulations on Job Duties of Teachers" specifies, "The higher education institutions with bachelor's degree-granting rights shall establish a Teacher Job Duty Evaluation Committee; and the higher education institutions without bachelor's degree-granting rights shall establish a Teacher Job Duty Evaluation Team." In the "Charters of Job Duty Evaluation Institutions for Teachers," the duties of the Job Duty Evaluation Committee for Teachers mainly include: review and approval of the qualifications of teaching assistants and lecturers; discussion of the qualifications of professors and associate professors and proposing opinions to the relevant departments in provinces, autonomous regions, municipalities, and state council ministries for approval; and the development of practical evaluation methods for the job duties and qualifications of teachers in the universities.

With the further promotion of the education reform, university autonomy has been gradually expanded, and at present most of the universities subordinated to the Ministry of Education have the right to evaluate teachers with high-level titles and professional technicians. Concerning the provisions governing the Title and Job Duty Evaluation Committee specified in the 7 university charters, their duties, rights, and responsibilities can be classified into the following two categories: review and approval of regulations and practical methods of title and job duty evaluation for teachers and professional technicians; and organization of the work of title and job duty evaluation for teachers and professional technicians. It is worth noting that in the charter of Wuhan University of Technology, the university has set up a Senior

Professional Skill and Job Duty Evaluation Committee that is responsible for evaluating the job duties of professors (researchers), while the right to evaluate the titles and job duties of associate professors and below are assigned to grassroots academic institutions; this approach to decentralization of power is worth learning from. In terms of staffing, the committee is composed of university leaders, persons in charge of relevant functional departments, and experts and professors in different disciplines, and the chairman of the Senior Professional Skill and Job Duty Evaluation Committee of Wuhan University of Technology is the president.

5. Other academic institutions in universities

In addition to the Academic Committee, Academic Degree Evaluation Committee, Professor Committee, and Title and Job Duty Evaluation Committee, there are some rare academic institutions specified in certain university charters, including Academic Ethics Committees, Discipline Establishment Committees, and Academic Evaluation Committees, all of which are established under the Academic Committee. In terms of legal bases, these three committees are based on the regulations of Article 11 in the "Regulations on Academic Committees."

Academic Ethics Committees have been established in only 3 universities: Shanghai Jiao Tong University, Shanghai University of Finance and Economics, and China University of Mining and Technology. They are responsible for reviewing the regulations on academic ethics, supervising and guiding the establishment of academic ethics, and academic style maintenance within the academic departments, accepting reports and complaints related to academic violations and organizing investigations and putting forward opinions.

So far, only one Discipline Establishment Committee has been established, in Shanghai University of Finance and Economics. It is responsible for discipline planning, establishment, and adjustment, as well as budgetary review for the various disciplines.

Only one Academic Evaluation Committee has been established, in China University of Mining and Technology. It is responsible for handling academic evaluations and academic controversies.

6. College-level academic institutions

Concerning the college-level academic institutions specified in 15 university charters, the Renmin University of China, Central China Normal University, Jilin University, Shanghai Jiao Tong University, and Southwest University have established college-level Academic Committees; ten universities, including Donghua University, Shanghai International Studies University, Central

**Table 6.2**　Academic institutions within colleges

| University | Academic Committee | Professor Committee | Talent Training Committee |
|---|---|---|---|
| Renmin University of China | √ | × | √ |
| Southeast University | ? | ? | ? |
| Donghua University | × | √ | × |
| Shanghai International Studies University | × | √ | × |
| Wuhan University of Technology | × | √ | × |
| Central China Normal University | √ | √ | × |
| Jilin University | √ | × | × |
| Shanghai Jiao Tong University | √ | × | × |
| Tongji University | × | √ | × |
| Sichuan University | × | √ | × |
| Northwest A&F University | × | √ | × |
| Northeast Normal University | × | √ | × |
| Shanghai University of Finance and Economics | × | √ | × |
| China University of Mining & Technology | × | √ | × |
| Southwest University | √ | × | × |

*Note:* The Renmin University of China has established Talent Training Committees at the university- and college-level, fulfilling the same functions as the university-level Teaching Committees (or Teaching Guidance Committees) at other universities. Southeast University allows its colleges to establish sub-academic institutions based on their own circumstances, but doesn't specify the names.

China Normal University, Tongji University, Sichuan University, Northwest Agriculture and Forestry University, Northeast Normal University, Shanghai University of Finance and Economics, and China University of Mining and Technology, have established college-level Professor Committees; and only the Renmin University of China has established a college-level Talent Training Committee. It is worth noting that Central China Normal University allows its colleges to establish Academic Committees or Professor Committees based on their own circumstances. Northeast Normal University has established college-level Professor Committees and department-level Academic Committees (Table 6.2).

## 6.3　Existing Problems in the Academic Governance System Specified in the Charters

Generally speaking, as a key part of the construction of the modern university system, all universities are still exploring how to improve the academic governance system. As can be seen from the statements in the above 15 university charters, although the universities have carefully considered and collected opinions, there are still some flaws, including the following obvious examples.

The legal basis has changed. In the process of university charter preparation and submission for approval, some legal bases have changed, and as a result, some of the university charters that have been reviewed and approved are inconsistent with the governing laws and need modification. For example the Ministry of Education issued the "Regulations on Academic Committees" after the submission and approval of the university charters prepared by the Renmin University of China, Southeast University, and Donghua University, so these three universities failed to designate the Academic Committees as the highest academic institutions in the universities.

Vague expressions. Some clauses are not precisely defined, and there are some loopholes. For example, the nature of the phrase "highest academic institution" is debatable, considering that the so-called "highest," the Academic Committee, either enjoys "the highest status" or exercises "the most power." If this is understood as meaning "the highest status," then the Academic Committee may become a mere figurehead. In addition, a small number of university charters specify among the duties and rights of the Academic Committee "to review and decide," but fail to clarify which matters are to be reviewed and which can be decided. These vague expressions may cause great confusion.

Empty statutory rights. The statutory decision-making rights of the Academic Committees are not detailed in the university charters. Although the "Regulations on Academic Committees" specifies that the Academic Committees will have rights including decision making, review, evaluation, and advising, through careful analysis of the clauses related to the duties and rights of the Academic Committees in 15 university charters, it can be seen that only the Academic Committees in Sichuan University, Southwest University, and Northeast Normal University have direct decision-making rights, and the matters involved in decision making are limited to the scope of academic evaluation, which coincide with the rights of "approval."

Unclear relationships. In most of the university charters, although the Academic Committees are defined as the highest academic institutions, the Academic Degree Evaluation Committees are not subsidiary bodies, so the relationship between the two is unclear. The typical case is the charter of the Beijing Normal University (not yet approved), which on the one hand specifies that the Academic Committee is the highest academic institution, and on the other hand specifies that the Academic Degree Evaluation Committee is the highest academic institution in handling affairs related to academic degrees. This will make it difficult for the two bodies to coordinate with each other, because both of them are "highest" academic institutions.

Loose flow and poor convergence. As specified in the university charters, the Academic Committees are responsible for reviewing and approving a number of academic affairs, but most of the university charters fail to clearly specify which departments receive and make final decisions (or how they make final decisions). In the leadership system of "CPC Party leadership and responsible president" widely used at present, the final opinions reached through review and approval by the Academic Committee shall be submitted to the CPC Party Committee (the standing committee or general committee) or in a President's Office Meeting for final confirmation. But in most of the university charters, the clauses related to the CPC Party Committee and president fail to clarify which person is responsible for confirming the matters submitted by the Academic Committees, which may lead to ineffective coordination between the academic departments and non-academic departments. As a result, the review and approval provided by the Academic Committee may not be effectively put into practice, rendering it effectively weak.

Some of these problems are a result of policies and regulations issued by the Ministry of Education, but most of them can be attributed to management defects in the universities. It is necessary to examine in-depth how the Ministry of Education can avoid flip-flops while leading universities in innovation, and leaving enough space for innovation in the universities. How can it improve the nature of charters being rigorously prepared, and stimulate the vitality of innovation within the universities?

## 6.4  An Academic Governance System Case Study: Northeast Normal University

Currently, Northeast Normal University has a wealth of experience in promoting "academic governance by professors."[8]

At the university level, Northeast Normal University currently has five committees: the Academic Committee, the Academic Degree Evaluation Committee, the Senate Committee, the Document Resource Committee, and the Budget Committee. These committees are parallel to each other, operating

---

[8]Source: Document submitted by the Northeast Normal University in the "211 Project" Modern University System Cum Charter Establishment Promotion Meeting Hosted by the Ministry of Education on May 22, 2014.

like independent parties. On the issue of the relationship between various types of committees and the president, the university adopts a governance approach involving "relatively isolated and limited constraints." In order to ensure the independent exercise of academic authority, highly specialized academic matters including academic resource allocation, academic evaluation, academic election and discipline establishment are not directly discussed in the President's Office Meeting, but these matters are instead decided by institutions such as the Academic Committee and then adopted in the President's Office Meeting, while the president has the right to ask for reconsideration.

At the college level, Northeast Normal University adopted a "Dean's responsibility system under the collective leadership of the Professor Committee," for which a term of membership is 3 years, with re-election for not more than two terms in the Professor Committee; the proportion of new members shall not be less than 1/3 after each election. The college leaders cannot be members according to clear provisions, but they can be invited to sit in on relevant meetings to clarify relevant affairs, although they have no voting rights. In order to clarify the relationship between the Joint Meeting and the Professor Committee, the charter of Northeast Normal University specifies that matters are to be decided by the Professor Committee in the Joint Meeting. Meanwhile, the charter also specifies the powers of the Professor Committee: "the important academic matters including talent training, discipline establishment, evaluation and reward of scientific research achievements, teacher appointment, talent introduction and team building, and international academic exchange and cooperation shall be submitted to the Professor Committee for review and decision making; and the Joint Meeting shall listen to the input of the Professor Committee on important matters including the overall development of the colleges, teaching and research organization, and resource allocation. In order to coordinate the relationships between academic departments and administrative departments, the charter of Northeast Normal University also specifies that the college CPC Party secretary shall sit in the Professor Committee and supervise its affairs; and the dean is responsible for developing the resolutions of the Professor Committee, but has the right to ask for one-time reconsideration of the matters discussed and determined in the Professor Committee.

There are three points worth noting in the case of the charter of the Northeast Normal University: 1) it establishes a relatively independent academic decision-making body less susceptible to interference at the university level, so that academic power predominates in the rights of utilizing key resources including talent, capital, and materials; 2) a management system with the Professor Committee at the core is established at the college level, and through empowerment of the Professor Committee within the Joint Meeting, "academic governance by professors" has institutional legitimacy; 3) a communication channel is established between the academic institutions and administrative institutions, in which the CPC Party secretary and administrators can sit in, and the responsible administers (the president and dean) can ask for reconsiderations, so that a mechanism of supervision and coordination between academic powers and administrative powers is formed. However, there are some problems with the academic governance system of Northeast Normal University: 1) the university-level academic institutions operate like independent parties, although the charter specifies that the Academic Committee is the highest academic institution, and there are insufficient communication and coordination channels, so the different committees may conflict with each other in management decision making, and they may have problems with the distribution of interests that may reduce their operating efficiency; 2) in the charter of Northeast Normal University, which has been reviewed and approved by the Ministry of Education, only the Academic Committee, Academic Degree Evaluation Committee, and the Senate Committee are specified in the charter, whereas the Document Resource Committee and Budget Committee are not. This may cause these two specific committees to lose institutional legitimacy and the right to speak on the distribution of key academic resources, such as budgets.

## 6.5  Policy Recommendations

At present, the preparation, review, and approval of university charters is in its final stages. More than 2,000 higher education institutions will have their charters approved in this year, and the work is complex, so errors are certainly possible. In the work of preparation, review and approval of university charters and the improvement of internal academic governance, universities and education authorities should carry out their duties to fulfill their critical responsibilities. It is necessary to build up a coherent system of linkages to resolve problems observed at present.

The higher education institutions, which prepare the charters, should adhere to academic autonomy, pursue dual-legitimacy, and prepare rigorous and practical provisions.

Adhering to academic autonomy means establishing an academic governance system capable of stand-alone operations and independent decision making. It is necessary to achieve autonomy at two levels: autonomy of university development against interventions of external forces, and autonomy of academic rights against administrative powers. To achieve academic autonomy, the idea is to realize "university administration by professors," and only when the teachers have "administration rights" can they guide the direction of the university development and reverse the "administrative achievement-oriented" direction, allowing talent training and truth-seeking to be the core functions and sacred missions of the university. Subject to the current political system and the external environment, "university administration by professors" has been dwarfed by "academic governance by professors." Nevertheless, it is necessary to hold to the principle of gradual improvement over the course of the preparation of the charters, to expand the breadth and strength of the professors' "scholarship rights," and lay a solid foundation for the realization of actual academic autonomy in the future.

Pursuing dual-legitimacy means balancing the formal legitimacy (Junjie, 2009) and substantive legitimacy (Junjie, 2010) of the charters and relevant academic governance regulations. So-called formal legitimacy means that the academic governance clauses specified in the charters shall be based on laws and coordinated with the current legal system in China, with the charters as the legal foundations. "Consistency with the laws" doesn't mean simply copying the laws, but rather entails making reference to the spirit of the law, and designing targeted, practical, and forward-looking clauses based on the actual circumstances of the universities. So-called substantive legitimacy means that the relevant academic governance systems shall be consistent with the values, ideals and beliefs of the academic community, and demonstrate autonomy in the charters. Academic autonomy should not stop at the spiritual and cultural level, but should further comprise the university system's ability to fulfill its functions of correction and guidance.

Preparing rigorous and practical provisions means preparing the charters with good logic and precise wording, which can be readily used as the basis of administrative proceedings and the object of judicial review. The university charters have the nature of administrative laws after approval, and as an independent legal entity, the university is bound to bear legal responsibility for the exercise of autonomous rights. Currently there are three common kinds of legal disputes involving universities: 1) litigation involving research funding issues; 2) litigation involving academic degrees; and 3) litigation involving academic integrity. All of these lawsuits involve academic governance. If the corresponding provisions of the university charters are not rigorous enough and relevant academic governance systems are not complete, then the universities will often be found groundless. Examples of lost cases in litigation include Tian Yong and Beijing University of Science and Technology, as well as the lawsuit between Liu Yanwen and the Academic Degree Evaluation Committee of Peking University (Zhongle, 2003).

The education authorities at all levels shall provide macro-direction, set appropriate approval standards, and make enough room for innovation.

Providing macro-direction. Direction is not administration, and rigid administration can only lead to repeating old mistakes. It is necessary to adopt measures including the preparation of laws and administrative regulations, and thus effectively guide universities down the right track. At present, the laws and regulations related to charter preparation and academic governance include the "Education Law," the "Higher Education Law," the "Methodology of Preparation of University Charters," and the "Regulations on Academic Committees," which have specified the autonomous rights enjoyed by the universities and shaped the basic framework of internal academic governance within the universities. In the university charter review and approval work in the future, the education authorities at all levels should fully respect the autonomy of universities in the preparation of their charters, and guide the university charter preparation work indirectly through seminars, exchanges and the issuing of references. It is not appropriate to enact new legislation to avoid mistakes and confusion.

Setting appropriate approval standards. The review and approval of university charters by the Ministry of Education is a necessary procedure to make the charters effective, and also a final procedure involving opportunities for modification and correction. Therefore, although there are a large number of charters to be approved, the competent authorities at all levels should check each one carefully and do their best to avoid significant changes after the charters are approved, so as to ensure the seriousness and stability of the charters. It is necessary to: 1) reasonably design approval procedures and reduce the possibilities of errors in the charters through review at three levels including universities, departments and specialists; 2) optimize the mix of experts, namely not only experts and scholars in the field of higher education, but also the professionals familiar with legal affairs and administrative directors familiar with policy directions, to improve the authoritative and scientific nature of the charter approval work.

Make enough room for innovation. Innovation is helpful to break through the shackles of the existing system and mindset, and through quantitative accumulation and qualitative changes, university governance and efficiency can be greatly improved. It is urgently important at this historical turning point to innovate in the charter preparation and system construction process, according to existing conditions, to remove administrative mistakes and construct a modern university system. But innovation means challenging the traditional powers and interests of vested groups, especially the education authorities at all levels. It is true that if we don't weaken and limit executive privileges of the education authorities, higher education institutions can never realize autonomous innovation. Therefore the education authorities at all levels shall modify their functions, remove unnecessary rights, get rid of unreasonable controls on universities, and make enough room for institutional innovation for universities.

## References

[1] Junjie, L. (2009). Discussion of Form Legitimacy of University Charters [J]. *Modern Education Management, 9.*
[2] Junjie, L. (2010). Substantive Legitimacy of University Charters [J]. *China Higher Education Research, 6.*
[3] Xiangju L. & Guangli, Z. (2004). A Legal Perspective of University Charters [J]. *Modern Education Science,11.*
[4] Yanghong, P. (2012). A Debate between "University administration by professors" and "Academic governance by professors" – Selection of Paths to Reform the Internal Governance Structures in China Universities [J]. *Tsinghua Journal of Education, 6.*
[5] Zhongle, Z. Higher Education and Administrative Proceedings [M]. Beijing: Peking University Press, 2003.

# 7

# The Construction and Development of the Chinese Open University

**Qinhua Zheng**

Beijing Normal University, Beijing, China

## 7.1 The Construction and Background of the Chinese Open University

Chinese Open Universities have developed from the Radio & TV Universities system. Central Radio & TV University was established in 1978. Since then, Radio & TV Universities have experienced a difficult growth process, moving from a position of weakness to strength and developing a Chinese form of distance education which includes Central Radio & TV University and 44 provincial Radio & TV Universities, 929 city-level Radio & TV Universities, 1,852 county-level Radio & TV Universities, 3,082 teaching centers and more than 60 thousand classes. This distance education network covers both urban and rural areas and serves more than 3.2 million students. Overall, the Chinese Radio & TV University system has served more than 7.5 million students and has also offered different types of non-academic continuing education (Yang, 2011). During its 30-year development as a complement to higher education, the Radio & TV University system has mainly relied on traditional higher education and played an important role within the Chinese higher education system. Especially at the beginning of China's reform and opening up, these schools effectively made up for the lack of higher education resources, offering opportunities to millions of young people aspiring to higher education. The Radio & TV University system is an important part of Chinese higher education and has made great contributions in promoting the popularization of higher education and education equality.

With the development of society and the economy, as well as continued development of education in China, the goal of offering adult education has effectively been accomplished. However, competition in the field of education is increasingly intense. Traditional universities are expanding in scale, vocational education is attracting more and more attention from the government, and related social forces and private education institutions are on the rise. A diversified investment pattern is gradually taking shape in education. All Radio & TV Universities are experiencing confusion about how to run their schools and undergoing structural reorganization.

For 30 years, open and flexible learning has been the goal of Radio & TV Universities, but two important factors have restricted their development. One is their lack of autonomy in running schools. Since their establishment, Radio & TV Universities have only been able to provide post-secondary education and undergraduate education; they have no power to apply majors, issue degree certificates, or offer graduate-level education. These restrictions hinder the development of Radio & TV Universities at the institutional level and also limit the universities' resilience in facing social demand. The other factor is that Radio & TV universities' system for running schools and developing learning content needs to change. With the development of society and changes in higher education, the definition of high-quality higher education has changed. Higher education must be more closely linked to local economic and social development, and more personalized and tailored to students' demand. Facing a demand for high-quality and personalized higher education with respect to learning content, teaching organization model, and other aspects, Radio & TV Universities have failed to adapt. It is therefore necessary to reform Chinese Radio & TV Universities using new institutional mechanisms (Zhang, 2013).

Lifelong education and universal education are two important approaches to education that emerged in the 20th century. With rapid social and technological transformation and progress, lifelong learning has become the inevitable tendency in the era of the knowledge economy, and international organizations as well as national and regional governments have all become involved in the improvement of a lifelong educational system and the construction of a learning city. An Open University has the advantages of covering urban and rural areas through a wide-ranging network system oriented to the whole society. Hence the Open University is an important breakthrough in the construction of a Chinese lifelong learning system that conforms to the requirements of Chinese government documents (Hao & Ji, 2012). In 2010, China released the "Outline of the National Program for

Long and Medium Term Educational Reform and Development Planning (2010–2020)," which addresses the development of lifelong learning practices in terms of basic concepts, approaches to fostering talent and development: "Measures shall be taken to speed up the construction of various types of learning organizations to initially build a learning society characterized with Learning for All and lifelong learning," including "erecting a 'flyover' of lifelong learning," "establishing the awareness of lifelong learning to lay a foundation for continuous development," as well as "pilot[ing] of building a mechanism of lifelong educational system." Not only is the Chinese Open University an important vehicle for the lifelong learning system, but it is also a subjective force and important element in this system.

In July 2012, the Chinese Ministry of Education approved establishing the Open University of China, Beijing Open University, and Shanghai Open University, modeled on Central Radio & TV University, Beijing Radio & TV University, and Shanghai Radio & TV University. The ceremony of the establishment of these three open universities was held in the Great Hall of People. After that, Jiangsu Open University, Yunnan Open University, and Guangdong Open University were approved as well. A national Open University and five provincial Open Universities have been established, which marks the start of Chinese Open Universities' construction and development. This was exciting news. Chinese Open University's "five plus one" pilot program is gradually taking shape. With the establishment of six Open Universities, the top layer design of Open Universities has become a focus of academic circles and society. In the process of developing a top level design, the Open University must first clearly define its substantial position, its functional differentiation from the general education system of national and provincial open education organizations, and its framework, and must clarify the special characteristics and standards of open education and its relationship with other educational institutions and social educational organizations, while also properly handling the deep inherent contradictions in open education, in order to guarantee its success. In the "Official Reply of Ministry of Education for Establishing Open University of China based on Central Radio & TV University" (2012), the Open University of China is described as a new university, which is directly under the Ministry of Education and will carry out distance education oriented to adults using modern information technology. The Open University of China offers both academic education and non-academic education. It will develop educational resources across the curriculum and make full use of traditional universities in order to promote resource sharing. In addition, it will advance the construction of an allied credit bank, establish a system for mutual recognition of academic achievements, and arrange the accumulation and transfer of credits so as to build a Lifelong Overpass. At the Conference on the Establishment of The Open University of China, Beijing Open University and Shanghai Open University, Yandong Liu, a government official who is responsible for Chinese education careers, emphasized that Open Universities, which are supported by modern information technology, should integrate and share high-quality education resources and innovate in teaching and learning models to establish an Open University with Chinese characteristics (Liu, 2012). Chinese Open Universities are intended to provide more flexible, fair, and open learning styles and multilevel, diverse learning services in order to make positive contributions to learning in society and empowerment in education and human resources. In this statement, we can see that administrative orientation and functions are both clarified. Chinese Open University adheres to both aspects of this function: education service and making a contribution to society.

Today, Chinese Open Universities' construction has moved past the initial stage. This article will tease out the background to their construction and their innovation practices in order to put forward policy suggestions for the next stage.

## 7.2 The Transformation Practices of Chinese Open Universities

The Open University of China and five regional Open Universities have been successful at the pilot project stage. These six Open Universities have explored reform to the setting of majors, teaching models, learning styles, credit bank construction, quality assurance, and so on. They have already established their own development model. The Chinese Open University "5 + 1"model is gradually taking shape.

1.  Exploring reform of the structure of majors, serving local social needs

Taking into consideration their different characteristics and local needs for social development, the six Open Universities set their development direction and adjusted their major structures. Forty-one new undergraduate majors were set up. The Open University of China has nineteen undergraduate majors; Beijing Open University, seven; Shanghai Open University, three; and Jiangsu Open University, Guangdong Open University, the Yunnan Open University, four each. As Beijing is one of the most rapidly developing cities in e-commerce in the world and many electrical businesses have their headquarters in Beijing, Beijing Open University set up an undergraduate major in the E-commerce Profession, aimed at promoting professional ability in e-commerce. In Beijing, many children with working parents have few chances to go to kindergarten. There is a serious shortage of teachers for preprimary education. Beijing Open University introduced a Preprimary Education major to train these teachers. One prominent characteristic of all the majors at Beijing Open University is the emphasis on professional competence. These majors are designed to prepare students to meet professional position requirements and are divided into three levels of four abilities.

Shanghai is a large, modern city, and there are many problems in the field of public safety. Shanghai Open University set up City Public Security Management major to address the high frequency of accidents and lack of city public security management talent in recent years. Jiangsu is a large agricultural province, and in order to cultivate talent for agricultural resource management and utilization, agricultural product safety management, and agricultural products sales and management, Jiangsu Open University set up the Agricultural Resources and Environment major. Guangdong province has a large population of immigrant workers, and labor disputes and the shortage of labor arbitration professionals are very serious problems. Guangdong Open University set up a Law major (Labor Disputes and Labor Arbitration) to address these labor disputes. In Yunnan Province, where the tourism industry is undergoing transformation and improvements, Yunnan Open University set up a Tourism Management major in order to satisfy the need to improve the quality and quantity of tourism professionals.

2. Carrying out education innovation and exploring Open University development

Integrating high quality educational resources, providing considerate support services, running an efficient education network, and establishing a platform for lifelong learning services is the common vision of the Open University system. These six open universities are actively studying international advanced experience and making innovations in different respects.

In resource construction: (1) they are oriented towards diverse learning subjects and enriching resource content. The Open University of China promotes different types of education, such as vocational education, community education, elderly education, and leisure education. Now the Community Education Website, the Student Village Official Website, and Sunshine Learning Website have been established in order to satisfy different learners' needs. The Open University of China is also planning to introduce some online courses from foreign universities and translate these courses for Chinese learners. Shanghai Open University has innovated by paying attention to leisure culture education, and has built some leisure culture resources.

In addition, Shanghai Open University has built the School of the Disabled, the School for Older Students, and the School for Women, to provide different education approaches and resources for different groups. Guangdong Open University provides learners with different types of resources links, such as networked open classes, National Excellent Courses, and so on, so that the learners can easily obtain different learning resources. In resource types: (2) At the end of 2012, The Open University of China launched a Five Minutes Course project, with five thousand micro-courses offered online so far. These courses cover more than 40 topics, such as Western economics, public economics, tea culture, photography, and so on. These courses are all developed to be accessed on mobile devices, and learners can spend their time taking these micro courses on the subway, on the bus, or in a queue. Confirming the tide and the development of Massive Online Open Courses (MOOC), the Open University of China has begun to take part in this movement. The initial plan is to launch 123 MOOCs in 2014, and 200 MOOCs should launch in the second stage next year. Beijing Open University has attended the iTunes U Project, to provide Chinese Open Educational Resources for learners all over the world, including 305 video resources for 15 courses in seven categories. In addition, Beijing Open University has transformed their methods of changing courses, encouraging students to take part in constructing the courses, in order to transform the courses from the default type into generative courses.

In exploring various cooperation channels: (3) Beijing Open University has explored cooperation with businesses in-depth. Modeled on the Software College of Beihang University and the Mobile and Cloud Computing training center of the Ministry of Industry, Beijing Open University has established a deep connection between industry capacity requirements and its subject knowledge system, in order to better and more accurately cultivate talent for companies and enterprises. In addition, some experts from traditional universities have been invited to construct the curricula and the resources in order to achieve a more professional orientation. Shanghai Open University has cooperated with traditional universities, a vocational college, and different adult education and training institutions. Yunnan Open University is trying to cooperate with enterprises to integrate education resources with local businesses. Jiangsu Open University has created Open Education TV programs with Jiangsu Education TV Station, highlighting knowledge, practice, service, and interests to enrich Jiangsu Open University's educational resources.

With respect to teaching reform, the Open University of China follows a "Six Networks" approach. The Six Networks are: online core curricula, online teachers' teams, online learning spaces, online learning evaluations, online credit banks, and an online Confucius institute. Making full use of modern information technology and the Internet, the Open University of China hopes to achieve a connection with provincial open universities, schools, and learning centers in order to build a convenient online campus. In addition, by using portable tablets, the Open University of China developed specialized mobile learning terminals, providing flexible learning to meet employed adults' diverse learning needs. Taking advantage of the online platform of the Open University of China and inviting experts and scholars from different fields, the University organizes different types of lectures for all members of society. During the pilot project, Beijing Open University developed a flexible curriculum in three stages. Beijing Open University uses a Moodle platform for curriculum management, which has functions for homework, discussion, tests, voting, and so on. This new platform can effectively display and present curriculum resources and is conducive to online interactive activities and promoting interactive online learning. Jiangsu Open University conducted research on the perception of mobile learning and which learning platforms students like most. They found that many distance learners like to use the WeChat app. Jiangsu Open University is trying to present course content and send notifications in this way. Yunnan Open University makes full use of Cloud Computing, 3G, and other modern information technologies to construct educational resources and promote teaching and learning interactivity to provide better Cloud services and distance learning support. Guangdong Open University has adopted a modular platform architecture by using Cloud computing technology to build a new networked teaching platform that integrates teaching, learning, management, services, and research.

With respect to the construction of teachers' teams, in accordance with their actual needs, the Open University of China provides regular and irregular training services for provincial Open Universities, in teaching, scientific research, management, distance learning support, and technical aspects. Beijing Open University established new departments, such as a Course Center, a Student Center, and a Teachers' Development Center. The goal of the Teachers' Development Center is to clarify the roles and responsibilities of Open University teachers, establish standards for tutors, and offer a training program to promote teacher development. The curricula of the seven new undergraduate majors have been transformed from preset into generative curricula. The Student Center tries to build a distance learner information model to provide learning support individually. Jiangsu Open University has also developed a Teachers' Development Center. The Teachers' Development Center is intended to confirm teachers' roles and responsibilities at the Open University, to develop the abilities of Open University tutors and its training scheme, and to promote teachers' professional development. Teachers at Jiangsu Open University are classified into three roles according to their different responsibilities: presiding teacher, academic tutor, and non-academic tutor. Presiding teachers are those who must take responsibility for setting curriculum objectives, designing and organizing curriculum content, and developing learning resources. Academic tutors are the experts who provide learning guidance, answer questions, and organize learning activities and evaluations. The academic teacher: student ratio is 1:35. Non-academic tutors provide administrative and information management services. The non-academic teacher: student ratio is 1:100.

Shanghai Open University upholds the spirit of open education by inviting outstanding individuals from different fields to give lectures or make speeches to students, which has expanded the scale of teaching and learning. Academic teachers receive training in modern technology and are encouraged to use new technology to carry out teaching activities and deliver learning resources.

3. Exploring credit bank construction, building a lifelong learning resource

Credit banks are an important measure and effective tool in the process of constructing a lifelong learning system and learning society. The Open University of China and the five provincial Open Universities have paid considerable attention to credit bank construction. Through credit certification, the integration of academic education and non-academic education will be achieved, and a lifelong learning resource will be built. Credit bank construction operates on three levels: national, regional, and institutional. The six Open Universities have taken its functional orientation and their own regional characteristics into consideration in the design of their credit banks, focusing on several different components.

The Open University of China is positioned at the national level to explore the establishment of a continuing education learning outcomes framework, which is the key to continuing education learning outcomes certification, accumulation, and conversion. This framework is the reference benchmark for different fields and different types of learning outcomes, and consists of learning outcomes levels, level descriptions, standards and specifications, categories and fields, and the rules and tools of credit accumulation and conversion. In addition, the Open University of China is actively exploring the mutual recognition of credits with the domestic and foreign universities, including continuing education schools and other educational institutions. On July 1, 2014, the Open University of China launched the construction of a credit bank information platform. This platform consists of three modules: the homepage, management system, and credit information database. The homepage presents the rules and processes as well as some dynamic information. The management system offers functions for establishing a lifelong learning account, conducting user management and business process management, and creating a course learning record. The credit information database is a database for storing all the credit for learned courses and achieved ability certifications. This is not an isolated platform for the Open University of China; the provincial universities all have the access and can collect the learners' information. This platform can be used to collect, process, and import credit information to the database. Though this credit bank information platform is in the pilot stage, it is still exploring new functions and aims to provide more and better services.

Beijing Open University first and foremost tries to build a qualifications framework, setting up conversion rules for vocational education and higher education. The qualification framework clarifies the name and main ability corresponding to each level, and carries out credit certification for industry and enterprise training courses. Now Beijing Open University has launched the e-commerce major's qualification framework, which ranges from a junior certification to a doctoral degree across seven different levels. The correspondence courses, from junior certification to high certification, are vocational education courses. The other certifications are post-undergraduate, undergraduate, and graduate level courses. Distance learners who have achieved certification in some field can transfer this into correspondence credit and store it in the credit bank. When they have passed the examinations and earned all the credits for a degree, a degree certification will be awarded.

Credit certification at Shanghai Open University is divided into three categories: academic courses, vocational training courses, and leisure culture courses. The credit bank enables students to convert credits between courses and majors and between vocational training and course credit, which provides direction and suggestions for other universities on the rules of credit conversion and ways of applying credit. Shanghai has opened a public account, setting up a municipal Courses Certification Center and Credit Certification Center so that it can provide certification and conversion between Shanghai Radio & TV University and spare-time universities, universities for employees, and self-study exams.

Jiangsu Open University is bidding for government support and concentrating on credit bank construction. Based on the platform of "Jiangsu Online Learning," Jiangsu Open University is exploring the construction of a credit bank and setting up a Credit Bank Management Center as a separate institution, which explains how to apply for credit. Anyone who registers with the Jiangsu Online Learning platform and takes courses at anytime, in any location, will earn credit that can be stored in the bank. At present, Jiangsu Open University has set some standards for credit transfer between courses and certifications; these majors include Modern Beauty, with three certification levels, a Beautician certificate (intermediate and above), a Makeup Artist certificate (elementary and intermediate), and an Office Applications Software Operator certificate (intermediate and above). In addition, Jiangsu Open University has founded the Teaching Support Alliance in order to lay a foundation for the implementation of learning outcomes (credits) certification and conversion between traditional high universities.

Yunnan Open University has set up a "Credit Conversion of Implementation Plan for National Vocational Certification and Yunnan Open University," which clarifies the relationships between 205 Vocational Certifications and 202 courses. In addition, Yunnan Open University has chosen more than 300 out of 3,232 courses in the cadre of online learning, Travel Online, and other learning platforms, to establish a cohesive relationship among the majors. This has provided experience in practice for constructing a credit bank for Yunnan Province.

Guangdong Open University has held several work symposiums on credit bank construction. In these symposiums, experts and scholars have discussed the basic ideas behind Guangdong lifelong education and credit bank construction work. Their goals, their basic orientation, the content of the credit bank, and its guiding principles and management system have been developed through these symposiums. Meanwhile, the implementation path, key linkages, and emphasis and difficulties involved in its construction are also being discussed.

4. Formulating quality assurance mechanisms, building an Open University brand

As new higher education institutions with new operating concepts and a new administrative model, Chinese Open Universities are vulnerable to questions about quality. So the development of quality is at the core of the university's mission, as the only path to responding to society's expectations and establishing the university's credibility. Quality assurance is a dynamic process, which is closely linked to teaching and management. Quality assurance can be divided into two dimensions: the external and internal quality assurance systems. External quality assurance generally involves competent education authorities, while internal quality assurance is provided by the university. The six Open Universities have different approaches to quality assurance development.

In order to further the development of the Open University of China and improve the quality of teaching and social service, the Open University of China has set up a Quality Assurance Committee, an Academic Degree Evaluation Committee, and an Academic Committee, studying and putting forward characteristically Chinese quality standards for Open Universities.

Beijing Open University has set up the Teaching Quality Assurance Committee and developed a Beijing Open University "Quality Assurance Manual." Cooperating with research teams from the School of Professional and Continuing Education at Hong Kong University and Beijing Normal University, Beijing Open University has set up a special quality assurance department. By researching and studying the quality assurance mechanisms of several famous universities, such as Open University UK, Phoenix University, and Hong Kong Open University, Beijing Open University has developed a series of quality assurance mechanisms that it has begun to pilot. In addition, Beijing Open University has received help and guidance from the Hong Kong qualifications agency secretariat and the Hong Kong council for academic and professional accreditation.

Shanghai Open University attaches great importance to quality assurance, using different approaches for academic education and non-academic education. As for academic education, the school headquarters provide unified guidance and management for schools in different counties, such as unified admissions, teaching materials, examinations, evaluation, and teachers' qualification identification. As for vocational training, Shanghai Open University uses the "university-school-learning center" three-level model. In this model, the school headquarters are responsible for developing vocational training projects at home or abroad; the project's training location is then arranged through the school, which is responsible for finishing the training task. School headquarters provide high-quality teachers and courses for each training center, providing regular training to improve teachers' abilities and monitoring and evaluating teaching quality for each learning center. For community education, the university likewise adopted a "university-school-learning center" three-level model. The municipal projects are conducted by school headquarters while the county projects and projects involving enterprises are conducted by schools, community schools, and vocational schools.

Jiangsu Open University emphasizes the "Integration of University, Platform, & System" that is an Open University, meaning a universal learning platform and lifelong learning support system (Peng, 2011). Jiangsu Open University is making efforts to achieve a connection between academic and non-academic education, between vocational and traditional education, and between pre- and post-vocational education. At first, the university focused on developing courses and digital learning resources and construction standards. Then, according to enrollment numbers, it organized different virtual classes. During the learning process, the teachers at Jiangsu Open University provided a wide range of student learning support services. Apart from the teachers' hard work on courses and learning support, monitoring course learning is the key to ensuring the quality of instruction. Implementing effective monitoring can ensure that each link in the teaching process is supervised effectively. Supervision covers two dimensions: the first is whether teachers are actively supervising virtual classes. This entails observing whether learners are spending their studying time on- or offline and whether they take part in the learning activities, and also observing the quality and quantity of online discussions. The other dimension is that of administrative work in the Teaching Quality Management Office.

In order to adapt and build a quality assurance system, Yunnan Open University set up an Academic Committee and a Professional Steering Committee to develop the "Construction Plan for the Quality Guarantee System of Yunnan Open University," along with more than ten other quality assurance regulations. In addition, it introduced a third party, the Michael Data Company, to participate in quality evaluation, and cooperated with the Yunnan Quality and Technical Supervision Department to guarantee teaching quality. Yunnan Open University has published a book entitled "The Construction of a Quality Assurance System for Yunnan Open University."

Guangdong Open University is striving to cooperate with certain distance learning research institutions, such as the Research Center on Distance Education at Beijing Normal University. With guidance from experts and scholars, Guangdong Open University has designed construction schemes for its curriculum and for student learning support, and has developed a teachers' development and quality assurance manual.

## 7.3  Challenges Facing the Chinese Open University

1. Academic and non-academic education: primary and secondary
   positioning

The documents of the Ministry of Education indicate that, "The Open University of China is a new type of university that is under the Ministry of Education and carries out distance learning for adults." "The Open University insists on academic education and non-academic education." From these documents, we can see that the Open University is a college that assigns equal importance to academic education and non-academic education. This is the first challenge facing the Open University. Out of academic and non-academic education, which is actually to be given priority? The six Open Universities have each answered this question in their own way.

As seen in the Table 7.1, the current administrative orientations of Chinese Open University and the other six Open Universities all treat the Ministry of Education's documents as guiding principles; that is, academic education and non-academic education are assigned the same importance, avoiding the problem of determining which is more important. These six Open Universities developed from Radio & TV Universities, and based on Radio & TV Universities' experience, there is no contradiction in pursuing academic education and non-academic education at the same time. However, as the operating rules of these two types of education are quite different, and the question of how to balance these two types of education is an important one. This is not to deny that the quality is satisfactory, especially for academic education. The goal of the Chinese Open Universities is to integrate high-quality education resources in China and around the world, to break out of the limitations of time and space placed on traditional education, and to solve the contradictions in higher education by expanding its scale, improving its quality, and reducing the

**Table 7.1** Administrative orientation

| University | Administrative Orientation |
| --- | --- |
| The Open University of China | Rapidly develop non-academic continuing education, steadily develop academic continuing education. |
| Beijing Open University | The fundamental purpose is to support society's learning and lifelong learning, providing diversified and multi-level academic education and non-academic education. |
| Shanghai Open University | A new Open University carrying on distance learning for adults, and providing academic education, vocational training and leisure culture education. |
| Jiangsu Open University | Academic education and non-academic education are assigned the same importance. |
| Guangdong Open University | Establish a new Open University serving lifelong education in a distinctive way, in line with international standards and domestic first-class education but using accessible language. |
| Yunnan Open University | Various approaches to administration, not only for open education, higher vocational education, adult education, secondary vocational education, and other forms of academic education, but also for cadre training, enterprise staff training, training for professional skill appraisal, and other forms of non-academic education. |

cost, so as to provide all members of society with more flexible and convenient ways of learning, and to offer more opportunities for students at different ages to receive the higher education.

As academic education and non-academic education are significantly different in orientation and in development mechanisms, it is a primary challenge for Open Universities to accurately determine which should be given priority.

From the view of administrative orientation, academic education and non-academic education are significantly different. Academic education emphasizes high quality and professionalism, and has stability and systemic characteristics intended to endure for a long time. The aim of academic education is to deliver knowledge to students. If learners complete relevant courses in a given major and pass the examinations, they can then earn degree certifications at different levels. Academic education is mainly provided by traditional universities, which are the primary mode of higher education in China. Non-academic education emphasizes rapid response to the market and the cultivation of talent to fulfill an immediate need, which demands flexibility and rapid adaptation to changing circumstances. The aim of non-academic education is to cultivate professional skills so that learners can easily find a job or better perform existing work. When students finish the learning process, they earn some type of certification in a particular ability rather than a degree. Non-academic education is offered by all kinds of educational institutions and enterprises. Traditional universities, at the top of the education development pyramid, take responsibility for cultivating high-quality talent. In general, traditional universities give academic education priority. A few universities founded on the basis of providing academic education also make full use of their high-quality resources (such as teachers and learning materials) to carry out non-academic education. The Open University, as a special university, needs to decide whether its primary task is to make a contribution to academic innovation and expand the potential of academic education with modern information technology and an open, flexible administrative orientation; or, instead, to be an open institution that primarily offers non-academic training. Academic education and non-academic education also differ greatly in development mechanisms. Academic education takes responsibility for systematically cultivating talent, not only in pursuit of human resource development for the government, society, the market, and individuals, but also as a leader in the arena of culture and values. In this context, academic education develops under the guidance of the government and through integrating social forces. Non-academic education is often provided in response to market demand,

and develops under the guidance of the market and through multi-channel, multi-body cooperation. These two orientations have important consequences for a school's development path. In the process of building a learning society, Chinese Open University responds to the demand for lifelong higher education and education equality, which is important to the development of China's lifelong learning system in the future. It is therefore necessary to decide whether the core function of Chinese Open Universities is to provide high-level academic education or to rapidly provide market-oriented, short-term, non-academic education.

Only by clarifying which is to be given priority in practice – academic or non-academic education – can the Chinese Open Universities innovate in higher education practice scientifically and effectively, providing high-quality education and ensuring the position of Open Universities in China's lifelong learning system.

2. The Relationship of Open Universities to the Radio & TV Universities

The five satellite Open Universities have been established for a year. From this experience, in practice, a realistic and concrete difficulty has emerged: the question of how to coordinate and deal with the relationship between the Open University and the Radio & TV University, which is also a challenge for the Chinese Open University.

The Ministry of Education has indicated that, compared with the Central Radio and TV University, the Open University of China is better and newer in eight respects: fundamental purpose, administrative mission, administrative orientation, administrative model, framework and operating mechanisms, technical support and learning model, objectives and quality assurance, and school properties and education level (Ministry of Education, 2012). With respect to framework and operating mechanisms, this document points out that the Open University of China system is more complete and has closer linkages between headquarters, branches, colleges, and learning centers. The Open University of China established its headquarters based on the approach of the Center of Radio & TV University. In accordance with the "voluntary, equal, and cooperative, wins" principle, the Open University of China has established branches modeled on the provincial Radio & TV Universities. According to the unified standards of the Open University of China, colleges and learning centers have been established modeled on the city- and county-level Radio & TV Universities.

The core institution in the Chinese Radio & TV Universities system is the Center Radio & TV University. The system has central, provincial, city, and county levels; that is, a four-tier vertical educational and management system. For many years of development and practice, the Chinese Radio & TV University system has formed a mature and effective educational model. Today, the construction of the Chinese Open Universities has had a huge impact on the original system. This impact is, first of all, a reflection of the relationship between the Open University of China and the other five provincial Open Universities. In the design of the Open University of China, the teaching and management institutions in different provinces are very important branches. These branches are modeled on the provincial Radio & TV Universities and are affiliated with the Open University of China. This means that the provincial Radio & TV Universities are taking on other roles than those of the branches of the Open University of China. The original relationship of superior and subordinate institutions still exists. When the Ministry of Education approved the establishments of the Shanghai, Beijing, Jiangsu, Yunnan, and Guangdong Open Universities, it established the new provincial universities and repealed the original mechanism of funding the Radio & TV Universities system at the same time. This is a new decision that will have a broad impact on the education system. As the new provincial universities, the five provincial Open Universities are led and managed by the different provinces of government; the expenditures on their development costs are also arranged by the different provinces. With the revocation of the mechanism for funding the five provincial Radio & TV Universities, existing relationships with the Open University of China do not exist. What are the new relationships between the Open University of China and the other five provincial Open Universities? Are the five provincial Open Universities still the branches of the Open University of China or are they new and independent provincial Universities? With the development of provincial Open Universities, we can imagine that more provincial Radio & TV Universities will be transformed into Open Universities, and that it will then be difficult for the Open Universities of China to follow the existing models for enrollment, administration, and management. Considering this, the transition from one type of university to another is a major test of government and institutional wisdom with respect to top design, and this is one of great systemic challenges facing Open Universities as well.

### 3. Capacity Building for the Open University

The key to developing the Chinese Open University is to satisfy the needs of all learners, to promote education equality, to avoid exam-oriented education, and to achieve quality education. The purpose of the Chinese Open University's development is to serve as an important support system for lifelong learning and to foster a learning society, which are important measures to serving national development and improving the national quality of life. At the "Conference on the Establishment of The Open University of China, Beijing Open University, and Shanghai Open University," Yandong Liu emphasized that when constructing an Open University, it was necessary to reform the administrative model and talent cultivation model, introducing strict, flexible teaching management and tolerant entry requirements as well as strict exit requirements (Liu, 2012). Registration, learning and exams at Open Universities should be more flexible and convenient, and the quality standards and guarantee system should improve the quality of higher education. Open Universities should push integration of education with information technology, improving teaching methods using learner-centered, web-based, self-directed learning and distance learning in combination with face-to-face teaching, creating a friendly digital learning environment. In addition, Open Universities should accelerate high-quality education resource-sharing, expanding high-quality resource types, the total amount of resources and coverage, in order to provide education services to people from all walks of life, especially grassroots learners. In addition, by cooperating with international universities and absorbing advanced concepts from successful experiences, Open Universities should continually improve their level of education and international influence. From these statements, we can see that the government has high expectations of the Open University system, hoping it can provide flexible and high-quality education services. To achieve this goal, Open Universities face a challenge in capacity-building. These challenges are primarily related to academic culture, system and support models, which comprise the capacity of the Open University.

With respect to academic culture, the Radio & TV Universities are oriented toward adult education and compensation for lack of access to traditional institutes of higher education. Although they have promoted the popularization of higher education and played an important role in Chinese higher education, there is no denying that their reputation for quality has been controversial in recent years. One of the controversies is over whether the Radio & TV University is a university or a higher education administrative system. In either case, the Radio & TV University has been transformed into an Open University, and its primary problem is determining how to achieve excellence. This means reforming the foundations of its academic culture from the old administrative system to a new type of university system: this, at least, is the approach expected by the government for Open University administrative reform and talent cultivation reform. The Open University's operating mechanisms should take open courses to be the core content, insisting on tolerant entry policies and strict exit policies. Tolerant entry policies aim to improve access to learning opportunities and high-quality learning resources, and strict exit policies enforce the commitment that participants will improve through education. Those earning a diploma will thus be more knowledgeable than they were before, and those who earn a degree will improve their capabilities.

In traditional Radio & TV Universities, the core features of academic education are the school year and credits, while Open Universities emphasize a strict, flexible teaching management system and a tolerant entry, strict exit learning system. At the external level, Open Universities also need a means of credit accumulation and conversion to build bridges to traditional universities, which means the design and implementation of a credit system, learning certification system, credit conversion system, student status management system, teaching service system, and so on. All of these reforms will be very difficult. Credit bank construction involves coordination with both traditional universities and institutes of vocational and other training. The goal is to achieve credit accumulation and conversion not just within each province, but also outside the province. Developing the relevant rules and learning outcome framework will be important but very difficult. How to design a student-centered learning and teaching system is another challenge facing the Open University.

The question of how to reform the student learning support model is also a challenge for the Chinese Open University. Distance education is student-centered and treats students as consumers, but students are not the real consumers, and nor are the consumers God. Sometimes there is a conflict between ensuring quality and providing individual services. Providing individual learning support doesn't mean lowering standards for learners. The Open University is trying to provide flexible and individual learning support from enrollment to graduation. As Open Universities undertake their academic and non-academic education mission, student learning support approaches should also change. One suggestion is to provide a wide range of ways to interact with students using modern information technology. Through the network management platform, teachers can supervise students' learning process and help solve problems in a timely fashion.

Since the establishment of the Chinese Open Universities, all six universities have placed emphasis on quality and made great efforts to guarantee quality, ranging from enrollment to the learning process. However, their reputation is not competitive. How they can improve their social standing and reputation is also a challenge.

On the whole, the Open University has a long way to go in its construction. To keep improving its construction capacity will be key to influencing actual outcomes for the Chinese Open University. However, improving capacity will require not only learning from the successful experience of Radio & TV Universities, but also continually innovating in the university's operating mechanisms in order to realize its strategic goals.

## 7.4 Conclusions

By breaking out of the barriers of time and space, being flexible and convenient, exercising resource sharing and enjoying systemic advantages, the Open University has already become the most popular and fastest growing form of education in China. In this new environment, Chinese Open Universities have achieved great innovation in their administrative models, talent cultivation models, evaluation models and service models. The Chinese Open Universities share certain general specifications, but they are differentiated in other respects. The general specification is primarily comprised of three dimensions: first, an Open University must have the basic attributes of a university. Second, the Chinese Open University must share its core identity with other Open Universities in the world. Third, different provincial open universities must be convergent in identity and practice. As for their individual characteristics, they have diverged in three ways: first, compared with traditional universities, Open Universities must seem unique. Second, compared with other Open Universities in the world, Chinese Open Universities must have certain ethnic characteristics. Third, the Open Universities in different provinces must each have their own locally oriented characteristics.

In the future, the Open University should give priority to academic education, while also promoting non-academic education. Academic education is the foundation of any university, and all universities in the world organize high-quality academic education. The Open University must first concentrate on pilot majors. Pilot majors should satisfy social needs, and majors' construction must be oriented to social demand. The purpose of cultivating courses must be to satisfy the need for jobs, and pay attention to students' self-directed learning abilities, problem solving abilities, and creative abilities. The content of courses must cover the latest relevant practical experience. Teaching must take on new methods and make full use of Web 2.0 tools in approach. Finally, curriculum development must employ a curriculum group organizational model, attaching importance to instructional design. Majors of the Open University must have its own characteristics to differentiate them from those at traditional universities. The non-academic education offered by the Open University must employ new models, exploring ways to encourage all of society to participate. The Open University should transform itself from being the provider of resources to being a resource organization institution, certification institution, and resource-sharing platform. Teaching at the Open University should give priority to academic education, and the social service mission of the Open University should give priority to non-academic education. A double-centre model should be used as the business model.

The most important thing is that the Open University must build an effective quality assurance system. Since its inception, Chinese distance education has emphasized quality, and the Ministry of Education and administrative institutions have made great efforts to strengthen quality management. However, quality problems have appeared, and their reputation for quality is worrisome. Although we often hear about encouraging success stories at internal meetings on higher distance education, there is widespread doubt on this point. In-depth research suggests that people are confusing quality management with quality assurance. Quality assurance entails all planned and systemic activities that make people believe that an entity can meet its respective quality requirements, implement activities within the quality system and confirm their implementation as required. Quality management entails activities intended to clarify quality policy, objectives, and responsibilities, by means of quality planning, quality control, quality assurance, and quality improvement within a quality system to achieve all management functions. Quality assurance aims to make people believe in quality, while quality management aims to achieve quality. Though these two things are closely related, they have different goals. Quality assurance rests on quality management, but quality management alone cannot convince the public that quality has been achieved. It is a pity that we never established a quality assurance system for the Open Universities system. Though some Radio & TV Universities and some network schools have established internal quality assurance, they never release this information to the public, so these assurances are still a type of quality management. As national higher education quality assurance has not been established, we suggest exploring institution-level quality assurance and releasing relevant information to the public, while accepting social third party assessment and regularly publishing its results, in order to promote the establishment of national higher education quality assurance.

In conclusion, the Chinese Open University faces several major challenges. The Chinese Open University should take its present situation into consideration and insist on its own administrative orientation in the direction of development, putting quality at core of its mission. Teaching work is given priority in academic education; social service is given priority in non-academic education. The Chinese Open University should make efforts to build a higher distance education quality assurance system so that it will achieve a high-quality status for Chinese higher distance education and be a first-class Open University.

## References

[1] Hao, K., Ji, M. (2012, July 30). Strategic initiatives of building national Lifelong Learning System—Making efforts to construct Chinese Characteristic Open University. *China Education Daily*.

[2] Introduction of Beijing Open University. [EB/OL]. Retrieved from http://www.bjou.edu.cn/web/about.jsp

[3] Introduction of Guangdong Open University. [EB/OL]. Retrieved from http://www.ougd.cn/

[4] Introduction of Guangdong Open University. [EB/OL]. Retrieved from http://www.ynou.cn/A10/A10001/A10001001/2011_07/29/1311904968742376.html

[5] Introduction of Jiangsu Open University. [EB/OL]. Retrieved from http://www.jsou.cn/read2-page.htm?id=4162

[6] Introduction of Shanghai Open University. [EB/OL]. Retrieved from http://www.shou.org.cn/Main/Content.aspx?i=1&j=0

[7] Introduction of The Open University of China. [EB/OL]. Retrieved from http://www.ouchn.edu.cn/Portal/Category/CategoryList.aspx?CategoryId=924df1f7-0cbb-414f-9abb-257ba7c97f50

[8] Liu, Y. (2012). Insist on facing all society, innovate mechanism; make efforts to establish Chinese character Open University. [EB/OL]. Retrieved from http://www.moe.gov.cn/publicfiles/business/htmlfiles/moe/moe_838/201208/140111.html.

[9] Ministry of Education. (2012). What are differences between The Open University and Central Radio & TV University? [EB/OL]. Retrieved from http://www.moe.edu.cn/publicfiles/business/htmlfiles/moe/s6073/201208/140343.html

[10] Official Reply of Ministry of Education for Establishing Open University of China based on Central Radio & TV University. (2012). [EB/OL]. Retrieved from http://www.moe.gov.cn/publicfiles/business/htmlfiles/moe/s181/201207/xxgk_138826.html.

[11] Peng, K. (2011). The Construction Ideas of Open University—The Implement Plan Based on Jiangsu Province. *Open Education Research*, 3, 10–17.

[12] Yang, Z. (2011). The history of Chinese Radio & TV University. [EB/OL]. Retrieved from http://dianda.china.com.cn/news/2011-05/11/content_4190773.html.

[13] Zhang, T. (2013). Push forward the connotation construction of Open University, to provide high quality educational services. *Learned Journal of Beijing Open University*, 2, 3–6.

# 8

# Reflections on the Development Patterns and Characteristics of the Chinese Credit Bank System

**Zhiying Nian**

Beijing Institute for the Learning Society, Beijing Normal University, Beijing, China

## Abstract

It has been almost ten years since the credit bank was implemented in practice in China. There are three comparatively established models: the pilot credit bank, the university-company collaboration, and the university-university collaboration. In terms of breadth, the practice of credit banking has developed on a large scale across the whole nation; in terms of depth, the construction of the credit bank still needs to be improved with respect to certificate framework, methods, institution, and mechanism. Thus, the next step should focus on the construction of a certificate framework, the clarification of management roles and participation, the design of a credit-transfer system, and the certification of informal and non-formal learning.

**Keywords:** Credit bank, achievement certification, continuing education, institutional innovation.

With the rapid development of the knowledge-based economy, the qualifications of all employees in China need improvement, which has become the core challenge to pursuing the Country with Rich Human Resources (CRHR) designation. The "National Education Reform and Development of Long-term Planning Programs (2010–2020)" drafted in 2010 clearly states that "constructing a lifelong learning 'highway bridge' means to promote the connections between all levels of education vertically and horizontally," "and launching the system of credits accumulation and transformation in continuing education will achieve confirmation and transition across different kinds of learning achievements," and this document proposes "to establish an academic credit bank system." In 2013, the "Decisions and Plans of the Party Central Committee for Deepening Reform Comprehensively" still stressed the importance of a plan "to pilot the credit transformation among institutions of higher education, vocational colleges, and tertiary institutions for adults for widening lifelong learning channels."

The practice of academic credit banking started with a collaboration between higher vocational colleges and enterprises in Beijing beginning in 2006. In June 2012, "The Theoretical and Practical Study on the Confirmation, Accumulation, and Transformation of Learning Achievements in National Continuing Education" was to launch, issued as the Project of Chinese Ministry of Education on the study of a national "academic credit bank system." This project is linked to the Open University of China, focusing on the framework, methodology, system, and mechanisms of the national "credit bank" policy and exploring it in practice with the construction of the Open University. From that point forward, the study on the construction of a credit bank moved forward from practical exploration into standardization.

## 8.1 Accelerated Social Development Stimulating Learning Demand

According to the "Major Figures of the 2000 National Population Census," 0.768 billion employees in China have relatively low levels of education, technical skills and training participation. In a science-oriented culture, from a knowledge and technical skills point of view, employees with junior secondary education and below make up 81.7% of the workforce, and those with senior secondary education and beyond occupy 18.3% (Hao, 2009). Table 8.1 shows that by 2009, the first group of employees accounted for 79.8%, while the second group had increased by 1.9%.

Table 8.1 shows that the education level of laborers has increased year by year, but there is still a huge gap compared with developed countries (Table 8.2). Up until 2006, employees with senior secondary education and beyond made up 18.53% of the workforce, a mere 0.23% increase compared with the year 2000. At this rate, the second group will have to wait 1,261.75 years to catch up with the average level in OECD countries, which is 69%.

From these stark contrasts, it can be seen that education levels in China urgently need to improve. In addition to the national education system, the main approach is to enhance the levels of all employees through recurrent

**Table 8.1** The constitution of all levels of education for employees in China (2002–2009) (Unit: %)

| Year | No Schooling or Education | Primary Education | Junior Secondary Education | Senior Secondary Education | College Education | University Education | Graduate Education |
|------|------|------|------|------|------|------|------|
| 2009 | 4.8 | 26.3 | 48.7 | 12.8 | 4.7 | 2.5 | 0.23 |
| 2008 | 5.3 | 27.4 | 47.7 | 12.7 | 4.4 | 2.3 | 0.21 |
| 2007 | 6.0 | 28.3 | 46.9 | 12.2 | 4.3 | 2.1 | 0.20 |
| 2006 | 6.7 | 29.9 | 44.9 | 11.9 | 4.3 | 2.1 | 0.23 |
| 2005 | 7.8 | 29.2 | 44.1 | 12.1 | 4.5 | 2.1 | 0.18 |
| 2004 | 6.2 | 27.4 | 45.8 | 13.4 | 5.0 | 2.1 | 0.13 |
| 2002 | 7.8 | 30.0 | 43.2 | 13.1 | 4.3 | 1.6 | 0.1 |

*Source:* The table was edited by the author, drawing from the China Labor Statistical Yearbook 2002–2010 and the China Population and Employment Statistics Yearbook 2002–2010.

**Table 8.2** Comparison of the level of education between China and developed countries (Unit: %)

| Country | Junior Secondary Education and Below | Senior Secondary Education | Higher Education |
|------|------|------|------|
| France | 33 | 41 | 35 |
| Japan | – | 60 | 41 |
| Korea | 23 | 44 | 32 |
| Britain | 31 | 39 | 30 |
| America | 13 | 48 | 39 |
| OECD member countries (average) | 31 | 42 | 27 |
| China | 81.5 | 11.9 | 6.63 |

*Source:* OECD. Education at a Glance, OECD Indicators 2008; China Population Statistics Yearbook 2007 and China Labor Statistical Yearbook 2007.

and continuing education. While academic education is in pressing need of restructuring overall, there is also a growing need for non-academic continuing education. In 2010, a total of 52,578,900 students signed up for non-academic education for adults (including those signed up for higher non-academic training for adults and intermediate vocational training for adults), accounting for 4.73% of the students of 15 years old and above. Beijing had the highest proportion of students in non-academic adult education and training out of the population of 15 years old and above, at 19.7%, followed by Shanghai (14.48%), Yunnan Province (11.00%), and Xinjiang Uygur Autonomous Region (10.30%) (Li, 2012).

From these figures, one can conclude that one of the inevitable requirements of social development is to provide people with education and training opportunities through diverse channels. Meanwhile, the perfection of relevant service and management mechanisms is indispensible to increasing participation. A credit bank will break the barrier between formal education and non-formal education, so that people can record their entire life experience and educational background in one transcript. Growth and achievements will stimulate their passion and impetus for learning. In this way, people will indeed realize self-development through lifelong learning, which in turn will promote the development of a learning society.

## 8.2　Practice Pattern of Credit Bank

Now Beijing, Shanghai, Zhejiang Province, Shaanxi Province, Jiangsu Province, Yunnan Province, and Shandong Province, as well as Beijing Xueyuan Road University Community and Guangzhou College Town, have actively explored the construction of a regional "credit bank," from which they have accumulated precious experience. A stable credit bank development system has come into its initial shape in China. The following section will introduce the overall development pattern of the credit bank through its three development modes: the credit bank pilot, university-company collaboration, and university-university collaboration.

1. Credit bank pilot projects

China officially initiated the all-round promotion of credit banks last year. On July 11, 2013, the first thirteen pilot academic achievement certification sub-centers (certification bodies) of the Open University of China were approved, symbolizing the official commencement of a certification service system for the Open University of China. According to the "Notice of Application for Academic Achievement Certification Sub-center (Certification Body) of the Open University of China (GKRZ [2013] No. 1)," the first pilot sub-centers were selected from the open universities involved in the project, the "Study on and Practice of Academic Achievement Certification, Accumulation, and Transfer Systems of National Continuing Education." Tianjin Broadcasting and Television University, Liaoning Broadcasting and Television University, Shenyang Broadcasting and Television University, Shanghai Broadcasting and Television University, Anhui Broadcasting and Television University, Jiangxi Broadcasting and Television University, Qingdao Broadcasting and Television University, Shenzhen Broadcasting and Television University, Xi'an Broadcasting and Television University, Gansu Broadcasting and Television University, the Open University of China Bayi College, Zhuhai Broadcasting and Television University, and Ningbo Broadcasting and Television

University Cixi College were selected ("First pilot academic achievement certification sub-centers (certification bodies) of OUC are approved", 2013). In the meantime, in line with the development of the project and credit bank construction at the Open University of China, the construction of the second pilot certification sub-centers (certification bodies) will commence at the appropriate time.

a. Full preparation of the Open University of China

Credit bank construction for the Open University of China is progressing in an orderly fashion. As of July 2012, the Open University of China (OUC) had cooperated with 26 regular colleges and university and 18 ministries, departments, and commissions (as well as industries and enterprises) in academic and non-academic education, including the Ministry of Human Resources and Social Security, the People's Bank of China, the National Bureau of Statistics, the National Health and Family Planning Commission, and the Beida Jade Bird Group. National vocational qualification certificates, certificates for professionals, and job training certificates were introduced into academic education to fill the gap between non-academic education and academic education. In this way, OUC accumulated experience in mutual credit recognition and transfer between non-academic and academic education. China Central Radio and TV University has implemented course exemption according to its pilot education program, launched in 1999, according to which 100% of courses at collaborative colleges and universities, 50% of courses at non-collaborative colleges and universities, 40% of self-taught courses and national certificates (such as the College English Test Band 4 and National Computer Rank Examination) are recognized. Beijing employee quality-oriented education programs and equivalent replacement of non-academic education courses by academic education courses in the Beijing "credit bank" program have also been put into practice. Especially in the "double-certificate" education cooperation program implemented in collaboration with the Ministry of Human Resources and Social Security, 25 national vocational qualification certificates were introduced and 25 higher vocational specialties provided by China Central Radio & TV University for open education, so as to implement the "Double-Certification Reform Program for Higher Vocational Specialties" and to promote the organic integration of academic education with non-academic certification education (Linshu, 2012). Through these programs, China Central Radio & TV University accumulated practical experience in credit recognition, accumulation, and transfer for continuing education, and made full preparations for the commencement of a "credit bank" in Beijing.

b. Initiation of credit banking for continuing education in Shanghai

Shanghai is the first city to establish a municipal credit bank in China, which provides an active reference point for the construction of credit banks in other provinces and municipalities. In July 2012, the Shanghai Municipal Education Commission issued an official document to open the bank and decided to appoint Shanghai Open University to establish the Credit Bank of Shanghai for Continuing Education (Shanghai Municipal Education Commission, 2012). On August 7 of the same year, the Commission announced the official establishment of the credit bank of Shanghai for continuing education. This digitalized information platform offers convenience to learners, who can visit the website of the credit bank at www.shcb.org.cn to browse a detailed introduction to the credit bank, review its operating procedures and the rules for credit transfer among collaborating universities, and obtain contacts at its branches. When the learner registers an account through the website, he or she can deposit, accredit, and transfer his or her credit via the information platform and keep an effective lifelong personal learning file. As of November 15, 2013, a total of 409,238 learners had opened accounts in the "Credit Bank of Shanghai for Continuing Education" and saved their personal learning files, recording more than 940,000 achievement records. Nearly 100,000 credits had been transferred online from non-academic education to academic education among different universities and vocational training centers, not including the new semester starting in the fall of 2013 ("More than 400,000 Registers at Credit Bank of Shanghai", 2013).

c. Credit banks in sight in other provinces and municipalities

The Yunnan Provincial Department of Education appointed Yunnan Open University to study and construct the credit bank of Yunnan and to explore credit recognition and transfer for its future continuing education program. In December 2012, Yunnan Open University formulated a general framework for the credit bank and associated proposals, including the credit transfer mechanism for 4 undergraduate programs and 23 professional training programs and the credit recognition and transfer proposal for the integration of intermediate and higher vocational education. Utilizing the pilot project "Open Talent Cultivation Mode Reform through Integration of Intermediate and Higher Vocational Education" as its entry point, Yunnan Open University stipulates that those participating in vocational training or obtaining vocational qualification certificates may convert their achievements or certificates to corresponding course credits upon certification, and that identical or similar courses in intermediate and higher vocational education can also be converted to corresponding credits upon certification. Now there are

34,146 students enrolled in the integrated intermediate and higher vocational education programs of Yunnan Open University who register and keep records at the credit bank of Yunnan. In order to extend the learning network, Yunnan Open University has set up open education offices in 16 prefectural and municipal education bureaus throughout Yunnan Province. The management of colleges and universities are hired part-time to assist in the promotion of lifelong education. Portions of local governments are responding actively to this program. From 2012 until to now, ten secondary open colleges have been established in eight prefectures and cities, including Qujing, Yuxi, Kunming, Zhaotong, Baoshan, Lincang, Lijiang, and Nujiang.

2. University-company collaboration

Some provinces and municipalities are not only actively engaged in the pilot construction of credit banks, but are also building collaboration bridges between universities and companies, so as to provide a high-quality platform for continuing education and on-the-job training for the senior talent in these companies.

University-company collaboration projects in higher vocational education commenced in 2006 in Beijing, to jointly try out the "credit bank program." In the beginning, only the Beijing Federation of Trade Union College collaborated with SINOPEC Beijing Yanshan Company. Now, such collaborations have been extended to six large enterprises, including Beijing Gas Group Co., Ltd. For the most part, front-line employees of pilot enterprises are admitted to the "credit bank program." Enterprise recommendation, the results of professional skill examination, and the results of higher education examination for self-taught adults are taken into consideration at the time of admission. Teaching management and teachers for foundation courses in the program are arranged by the Beijing Federation of Trade Union College. Collaborating companies offer teaching and experiment bases and facilities for practical training. "Double-qualified" technicians are selected to teach professional courses. The Beijing Federation of Trade Union College also utilizes the distance education resources of TV universities to facilitate dispersed learning, team learning, and independent study. Students can convert technical training from companies and certificates obtained to corresponding credits. When they meet the requirements for credits, they will obtain a college diploma recognized by the state. In 2011, there were 5 pilot enterprises involved in the program and 309 students had been admitted, including 29 from SINOPEC Beijing Yanshan Company, 79 from BAIC Foton Motor Co., Ltd., 73 from

Beijing Gas Group Co., Ltd., 42 from Beijing Hyundai Motor Company, and 96 from BBMG Group Co., Ltd. (""*Credit Bank Program*" for Front-line Employees", 2012).

Yunnan Open University is collaborating with local governments, colleges and universities, industries, and enterprises through "four supporting unions" to integrate resources. It is focused on cultivating talents with local characteristics and promoting the construction of learning enterprises. The "four unions" have become an important source of support in promoting the construction of a lifelong education system and learning society. On December 19, 2011, Yunnan Open University signed and entered into cooperative agreements with representatives of Zhaotong Municipal Government, Lincang Municipal Government, Yunnan University of Nationalities, Yunnan University of TCM, the China Council for the Promotion of International Trade Yunnan Branch, and First Machinery Co., Ltd. and Second Machinery Co., Ltd. of KSEC to establish supporting unions (Run Yunnan Open University, 2011).

### 3. University-university collaboration

In China, universities collaborate to construct credit banks, primarily through two channels: credit recognition and transfer for academic education to regular higher education and credit recognition and transfer to higher continuing education. Beijing Xueyuan Road Community is a typical representative of the former. It is an innovation in the management mechanisms of higher education, as regular university and college members share high-quality resources and provide students with a more flexible learning style. The latter channel, which is the focus of this thesis, is an innovative and diverse learning approach by which the higher continuing education institutions provide in-service employees and the public with more convenient opportunities for further education.

Some provinces and municipalities have explored their own development paths in practice. For example, Guangzhou Province and Shaanxi Province have launched university-university collaborations for the construction of credit banks.

In Guangdong Province, university joint education has been adopted to promote the development of higher continuing education. This approach can be used effectively to gather advantageous resources and make up for the weaknesses of a university in school administration, such as sources of students, teacher resources, teaching resources, and geographical locations. In their collaboration, universities utilize modular design, a mutual credit recognition system, and course enrollment cards, dividing courses into common courses, foundation courses, intermediate courses, and advanced courses. A flexible schooling system is also applied. Students can graduate early if they obtain the required credits. In the meantime, students are allowed and encouraged to study as part-time students, to combine learning with working, and to complete their study by stages. Students will be granted graduation certificates when they obtain the required credits within a specified period of time (Weiqi, Zhonglin, Zhenmin, & Xin, 2012). Universities and colleges located in the same region optimize their course system, so that credits that students obtain by taking the same courses at any university or teaching school are recognized. The adoption of the mutual credit recognition system impels course reform and construction in Guangdong Province. Now, Guangdong Province is still exploring ways to combine higher continuing academic education with non-academic education.

On May 26, 2012, 17 universities and colleges in Shaanxi Province signed their names to the "Convention on Credit Bank of Shaanxi Province for Higher Continuing Education," which means that credits are mutually recognized, accumulated, and transferred among these universities and colleges for higher continuing education. These first 17 universities and colleges were Xi'an Jiaotong University, Northwestern Polytechnical University, Northwest A&F University, Xidian University, Shaanxi Normal University, Chang'an University, Northwest University, Xi'an University of Technology, Xi'an University of Architecture and Technology, Shaanxi University of Science and Technology, Xi'an University of Science and Technology, Xi'an Shiyou University, Yan'an University, Xi'an Technological University, Xi'an International Studies University, Xi'an University of Posts and Communications, and Shaanxi Radio & TV University ("17 Universities in Shaanxi Cofounded "Credit Bank" Yesterday", 2012). Students enrolled for continuing education may select study contents, time, and place at different universities and colleges at their own discretion. Credits can be accumulated and transferred among these universities and colleges. When the "credit bank" is set up, these universities and colleges will unify their standards with respect to admission qualifications, curriculum, and teaching quality for different administrative approaches to higher continuing education. Meanwhile, they will promote the inter-university and inter-specialty sharing of teaching resources, so that high-quality resources are available at any university at any time.

## 8.3 The Construction of the Shanghai Academic Credit Transfer and Accumulation Bank for Lifelong Education

On July 24, 2012, the Shanghai Learning Society Construction and Lifelong Education Promotion Meeting was held. At the meeting, Mr. Yin Yicui, Deputy Secretary of Shanghai Municipal Committee of CPC, unveiled the nameplate of the Shanghai Academic Credit Transfer and Accumulation Bank for Lifelong Education, symbolizing that the first provincial and municipal-level credit bank in China was officially being put into operation. One year later, a mature organization management mechanism was set up in the Shanghai Academic Credit Transfer and Accumulation Bank for Lifelong Education, and credit certification and transfers were carried out smoothly thanks to clear positioning. In the following section, the author will introduce the development of the Shanghai Academic Credit Transfer and Accumulation Bank for Lifelong Education in detail, covering its targets and positioning as well as its organization, standard system, and credit certification and transfer system.

1. Positioning

In line with the policy documents issued by the Shanghai Municipal Education Commission, the Shanghai Academic Credit Transfer and Accumulation Bank for Lifelong Education specifies its service objects and functions. "Notice of Shanghai Municipal Education Commission on Establishment of Shanghai Academic Credit Transfer and Accumulation Bank for Lifelong Education (HJWZ [2012] No. 6)" points out that "The Shanghai Academic Credit Transfer and Accumulation Bank for Lifelong Education, an important part of the Shanghai lifelong education system, serves all Shanghai citizens as a platform for recognition, accumulation, and transfer of academic achieve-ments in continuing education, a platform for learning ability certification for citizens, and a platform that records the academic achievements of citizens engaged in lifelong education." The notice also specifies the management and operation mechanisms of the Shanghai Academic Credit Transfer and Accu-mulation Bank for Lifelong Education as follows: "The Shanghai Academic Credit Transfer and Accumulation Bank for Lifelong Education is sponsored and managed by the Shanghai Municipal Education Commission under the direction of the Shanghai Learning Society Construction and Lifelong Edu-cation Promotion Commission. Shanghai Open University is appointed to be responsible for specific administration of the Bank."

It can be inferred from the above text that the Shanghai Academic Credit Transfer and Accumulation Bank for Lifelong Education is an academic achievement certification management center and a credit transfer platform that is completely oriented to learners in Shanghai and primarily used for credit recognition, accumulation, and transfer for lifelong education. It is set up with an objective of building a "viaduct" for lifelong education and promoting the construction of a lifelong education system and learning society in Shanghai.

In practice, the Shanghai Academic Credit Transfer and Accumulation Bank for Lifelong Education operates in line with the following four principles. First, the Bank is positioned to be a municipal credit bank that is an integral part of the lifelong education system of Shanghai, a vital source of support in the construction of a learning society, a significant terminal of the lifelong study "viaduct," and an important platform for IT services. Although Shanghai Open University is appointed to operate the Bank, it is not a credit bank of any particular university. Second, the Bank is focused on continuing education in the preliminary stage. According to the National and Shanghai municipal "Outlines for Medium and Long-term Education Reform and Development" that propose the exploratory application of a credit bank system to continuing education, the Shanghai Academic Credit Transfer and Accumulation Bank for Lifelong Education serves mainly learners and institutions of continuing education instead of students of regular universities and colleges. Third, the Bank is used to create a "viaduct" and serves as a "terminal" for lifelong education. It is a management and service platform that is mainly responsible for the management of academic achievement certification and credit transfer, rather than certification and issuing graduation certificates as the credit banks in ROK do. Fourth, the government is responsible for the construction of the bank. Shanghai Municipal Education Commission is responsible for the construction of the bank management center, while district and county bureaus of education are responsible for the construction of the branches of the bank.

2. Organization

The Shanghai Academic Credit Transfer and Accumulation Bank for Lifelong Education consists of administration organs and operation organs in terms of organization. The former include the Management Committee, Management Center, Expert Committee, and expert panels; while the latter include bank branches and outlets in universities (Figure 8.1).

The Management Committee, the leading organ of the bank, is responsible for macro-direction and decision making for the construction and operation of the bank. The Management Center is the actual administrative organ that is responsible for the construction, operation, and management of

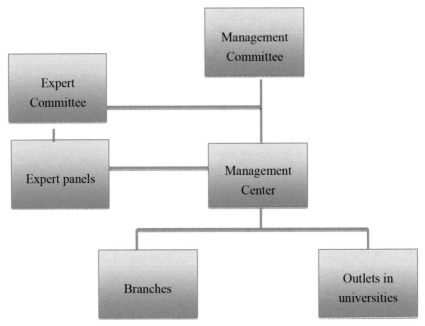

**Figure 8.1**   Organizations in the Shanghai academic credit transfer and accumulation bank for lifelong education.

the bank. Under the Management Center are the Academic Education Certification Department, Vocational Training Certification Department, Cultural and Leisure Education Certification Department, Information Service Department, and Comprehensive Management Department. The Management Center is located in Shanghai Open University. The Expert Committee is the consulting and direction organ that provides directions on development planning and the bank's operation approach. Expert panels are set up by disciplines, which are responsible for the formulation of credit recognition standards and the specific work in relation to credit recognition.

The operation system of the bank consists of branches and outlets in universities. Branches of the bank are set up according to administrative divisions in Shanghai. Currently, there are 21 branches in 17 districts. The branches are oriented toward society and responsible for the registration of learners at the bank, preliminary review for credit recognition, and consulting and publicity for the bank. "HJWZ [2012] No. 6 Documents" specifies that district and county bureaus of education are responsible for the construction of bank branches in their districts and counties, and they shall provide inputs to the branches in terms of manpower, equipment, facilities, and operating funds ("Notice of Shanghai Municipal Education Commission on Establishment of Shanghai Academic Credit Transfer and Accumulation Bank for Lifelong Education", 2012). Citizens may apply for registration online at the website of the bank and handle the specific formalities at nearby branches with their ID cards.

Outlets in the universities are oriented toward college and university students and are responsible for registration at the bank and credit recognition and transfer for students of the university, as well as consulting and publicity for the bank. So far, the bank has set up outlets in the continuing (adult) education schools of regular universities and other adult colleges in Shanghai. As of 2013, the Shanghai Academic Credit Transfer and Accumulation Bank for Lifelong Education had set up outlets in the continuing education or adult education schools of all regular universities, higher vocational colleges, and professional colleges, as well as independent adult colleges in Shanghai, including universities included in "985 project" such as Fudan University, Shanghai Jiaotong University, Tongji University, and East China Normal University, and all universities that are included in "211 project" and offer continuing education and adult education (Shanghai Academic Credit Transfer and Accumulation Bank for Lifelong Education, 2013). There were then 68 outlets in total.

### 3. Standard system and credit certification and transfer system

In the Shanghai Academic Credit Transfer and Accumulation Bank for Lifelong Education, credits are categorized into those for academic education, vocational training, and cultural and leisure education (community education and education for senior citizens). Credits for some forms of vocational training can be converted to those of academic education as per certain standards. In terms of recognition standards, credits for academic education will be recognized according to the standards of national education; credits for vocational training will be recognized according to the non-academic certificates and training programs certified by the bank; while credits for cultural and leisure education will be recognized according to the educational and learning programs certified and approved by the bank.

Based on the standards above, credits can be certified and transferred among bank branches and outlets in universities. The transfer of credits of academic education involves correspondence between courses and majors, and between courses and non-academic certificates obtained via vocational training. Such correspondence provides guidance and reference for bank outlets in universities to formulate provisions for credit transfer and applications of learners for credit transfer.

For now, the Shanghai Academic Credit Transfer and Accumulation Bank for Lifelong Education is focused on continuing education (Feilong, 2013). It is intended to connect and integrate education institutions and resources in the field of continuing education and realize communication between universities and society in terms of education of various kinds. According to the "12th Five-year Plan," by the end of 2015, Shanghai will complete the formulation of specialty standards for the credit bank, which will cover most specialties of adult education; a complete vocational training certificate system and the curriculum system for cultural and leisure education will come into being; and credit transfer will be effectively realized among universities offering adult academic education and open universities as well as between academic education and non-academic education, such as vocational training (People.com.cn, 2011).

## 8.4 Development Characteristics of the Credit Bank System in China

The construction of credit banks is an important institutional innovation for China's progress towards becoming a learning society (Zunmin, 2012). An institutional innovation of this kind will encounter many problems in policy, laws and regulations, mechanisms, and institutions. The credit bank system can be established through a comprehensive survey of the general situation and coordinated design, instead of through the resolution of certain problems before others. Based on the broad practical experience of the provinces, municipalities, and autonomous regions in credit bank construction, the development of the credit bank system has the following features: first, credit bank construction is promoted under policy guidance while local governments are slow in practice; second, practice is diversified in forms while credit bank construction lacks systematic national coordination; and third, resource integration is promoted under trans-institutional leadership while the cost of communication increases due to the need to coordinate multiple parties.

1. Credit bank construction is promoted under policy guidance while local governments are slow in practice

Government guiding strategy has played a notable role in the construction of the credit bank system in China. From the "National Outline for Medium and Long-term Education Reform and Development" to "Several Opinions on Accelerating Development of Continuing Education" and "Working Guide 2012 and Decision of the CCCPC on Some Major Issues Concerning Comprehensively Deepening the Reform," credit banks were first mentioned in national overall decisions and discussed as a special topic of continuing education, then became an emphasis in annual work plans, and are now a major issue. Such change shows that national policy on credit banks is becoming increasingly specific and focused.

In contrast to policy promotion, in practice, credit banking is slow and difficult work. According to the pilot institutions that implement credit banking, their achievements have been fruitful. Generally speaking, however, their explorations have been limited, and no complete and stable development pattern has taken shape. Owing to the continuing education system and other reasons, there is no perfect academic achievement certification mechanism or organization in China. Detailed rules for implementation and operable measures are scarce. Therefore, China is unable to fundamentally build up a complete "credit transfer system." Though some provinces and municipalities started credit bank construction early, the problems they've encountered in practice remain. The Fujian Provincial Department of Education explored and set up a "credit bank" and "curriculum supermarket" in 2009, and there still is no perfect credit bank system ("Fujian Province Will Establish "Credit Bank" and "Curriculum Supermarket"", 2009). In sum, it is arduous and onerous to put theories into practice. The success of a few provinces and municipalities has helped build confidence. The precious experiences of Shanghai, Beijing and Cixi comprise the successful foundations for construction of credit banks in the future.

2. Practice is diversified in form, while credit bank construction lacks systematic national coordination

Credit banks can be established by the central government, by local governments, or by institutions. In Great Britain and Australia, credit banks are constructed by central governments; in Hong Kong and Shanghai, credit banks are established by local governments. In Great Britain, credit banks are a part of the organization of open universities. Credit banks are varied in application and scope. Some are used for higher education, such as the Bologna Process launched by the EU, and the academic credit accumulation and transfer system adopted in Great Britain. Some are oriented toward vocational education, such as the EU's Copenhagen Process. Some are used for credit transfer between higher education and vocational education, such as those adopted in Australia, the US, Canada, Hong Kong, and Shanghai (Ying, Runzhi, & Ronghuai, 2012).

The practice forms of credit banks vary from province to province, such as the provincial (municipal) credit banks adopted in Shanghai and Jiangsu Province; city credit banks in Cixi, Zhejiang Province, and Nanjing; and regional "credit banks" such as the Beijing Federation of Trade Union, Xueyuan Road Community, Shaanxi University Collaboration Convention, Yunnan Union to Support Open Education, and Guangzhou College Town. Each of these models has accumulated precious experience in the exploration of credit bank construction. Though China has a vast territory and regional differences are enormous, researchers and practitioners need to think further about whether a national academic credit recognition, evaluation, and transfer system (including overall design, standards, and technical schemes) could be established on the basis of pilot credit banks in some regions and educational institutions.

3. Resource integration is promoted under trans-institutional leadership while the cost of communication increases due to the need for coordination among multiple parties

Credit banks have been constructed to break down barriers among different educational systems; to set up smooth communication systems; and to establish "viaducts" that enable mutual communication among regulators on education, vocational education, and continuing education. For the purpose of effective communication involving different educational systems, administrative barriers must be broken. For this reason, leading groups of the credit banks are effectively trans-department syntheses concerning commissions, offices, and departments. This method breaks the limitations associated with divisional management but increases communication costs.

Institutional innovations will inevitably encounter bottlenecks in their development. How to find the breakthrough to solve these problems, however, is the key to future development.

## References

[1] 17 Universities in Shaanxi Co-founded "Credit Bank" Yesterday [N]. (2012, May 27). *Xi'an Evening*. Retrieved from http://news.hsw.cn/system/2012/05/27/051329981.shtml.

[2] "Credit Bank Program" for Front-line Employees. (2012). Retrieved from http://www.bjeea.cn/html/ksb/chengzhaozhuanban/2012/0328/41890.html.

[3] First pilot academic achievement certification sub-centers (certification bodies) of OUC are approved [N]. (2013, July 29). OUC On-line. Retrieved from http://dianda.china.com.cn/news/2013-07/29/content_6164026.htm

[4] Fujian Province Will Establish "Credit Bank" and "Curriculum Supermarket" [N]. (2009, February 6). *Fuzhou Evening*. Retrieved from http://www.eol.cn/fujian_5295/20090206/t20090206_357082.shtml.

[5] Hao, K. (2009). Continuing Education and Institutional Construction for Chinese Employees [J]. *World Education Information*, 1, 32–35, 41.

[6] Li, L. et al. (2012). Report on Development of Continuing Education in China 2012 [M]. Beijing: Educational Science Publishing House.

[7] Li, L.(2012). Promoting Strategic Transformation of TV Universities to Open Universities Utilizing the Study on and Practice of National "Credit Bank" System. Retrieved from http://dianda.china.com.cn/news/2012-07/09/content_5146754.htm

[8] Luo, W., Wei, Z., Hu, Z., and Ding, X. (2012). *Strategy and Policy: Research on Development of Continuing Education in Guangdong Province* [M]. Guangzhou: Guangdong Higher Education Publishing House.

[9] More than 400,000 Register at Credit Bank of Shanghai [N]. (2013, November 28). *Jiefang Daily*. Retrieved from http://newspaper.jfdaily.com/jfrb/html/2013-11/28/content_1120926.htm.

[10] Notice of Shanghai Municipal Education Commission on Establishment of Shanghai Academic Credit Transfer and Accumulation Bank for Lifelong Education. (2012). Retrieved from http://www.shanghai.gov.cn/shanghai/node2314/node2319/node12344/u26ai32873.html.

[11] Peng, F. (2013).*Theory and Technology of the Credit Bank for Lifelong Education System* [M]. Beijing: Higher Education Publishing House.

[12] People.com.cn. (2011). Shanghai Academic Credit Transfer and Accumulation Bank for Lifelong Education Put to Trial Operation [N]. Retrieved from http://edu.people.com.cn/GB/16533333.html.

[13] Run Yunnan Open University, Development Lifelong Education [N]. (2011). Retrieved from http://yn.yunnan.cn/html/2011-12/19/content_1961043_2.htm.

[14] Shanghai Academic Credit Transfer and Accumulation Bank for Lifelong Education. (2013). Retrieved from http://www.shcb.net.cn/.

[15] Shanghai Municipal Education Commission. (2012). Notice on Establishment of the Credit Bank of Shanghai for Continuing Education (HJWZ [2012] No. 6). [EB/OL]. Retrieved from http://www.shanghai.gov.cn/shanghai/node2314/node2319/node12344/u26ai32873.html.

[16] Wang, Y., Zhang, R., & Huang, R. (2012). Practice and Analysis of Credit Bank Construction with an International Perspective. [J]. *Distance Education in China, 6,* 47–54.

[17] Wu, Z. (2012, October 15). Can Credit Bank Be Constructed by Self-educated Examination Institutions? [N]. *China Education Daily*. Retrieved from http://edu.ifeng.com/gundong/detail_2012_10/15/18259615_0.shtml.

[18] Yunnan Open University: a new university for lifelong education [N]. (2012, December 25). *Yunnan Daily*. Retrieved from http://yndaily.yunnan.cn/html/2012-12/25/content_657750.htm?div=-1.

# About Editors and Contributors

**Dayong Yuan** is a researcher from the Institute of Vocational and Adult Education (IVAE) at Beijing Academy of Educational Sciences (BAES); he is the CONFINTEA Scholar at UNESCO Institute for Lifelong Learning. His research mainly focuses on lifelong vocational and adult education, international, and comparative education. Currently he is working on a number of research projects about vocational education in Beijing, and at the same time, he provides consulting and academic services to the Beijing Municipal Education of Commission, as well as the vocational schools and colleges in Beijing. He is also responsible for international exchanges and cooperation in his institution. He has published an English language book entitled *Towards The Learning City of Beijing: A review of the contribution made by the different education sectors (Glasgow Caledonian University)*, and many articles in Chinese journals.

**Li Chen**, was born in May 1964 in Tianjin. She holds a doctorate and serves as a professor and doctoral candidate supervisor at BNU. She is also a BNU Vice President, an adjunct Director of the Office of Development Planning and Discipline Development and Executive Dean of the Beijing Institute for the Learning Society. She presides over arts scientific research and information systems development. Her areas of responsibility are the Social Sciences Administration Office, the Center of Information & Network Technology and the Journal of Beijing Normal University (Arts).

Li Chen studied at the Nankai Middle School, in the Nankai district of Tianjin, and later at Beijing Normal University. She has worked as League Secretary of BNU's Department of Radio Electronics, as Vice Dean of BNU's College of Information Science and Technology, as Vice President of the BNU Trade Union, as Party Secretary for the School of Educational Technology at BNU, and as Director of the Office of Development Planning and Discipline Development. At one point she worked as a volunteer teacher at Beijing Changping Teachers Training School. She has also been a visiting scholar at the Open University, UK, and at the Open University of Hong Kong.

She serves now as a member of the University Distance Education Professional Committee of the China Association for Educational Technology, as Director of the Primary and Secondary School Computer Education Professional Committee of the Chinese Society of Education, as Consultant to the Mechanism Development Group for Lifelong Education System, National Education Advisory Committee and as a Member of the Academic Committee of the China Strategy Society. She is also a member of the Expert Consultation Committee of the Beijing Modern School Network Education Institute.

Li Chen's main courses include: remote education foundation (undergraduate), remote education professional discussion – problems and trends (Master's degree candidate level), distance education theory and practice (at the doctoral degree candidate level). Her main research fields are teaching interaction in distance learning, distance education quality assurance and the evaluation of learning city. Among other distinctions: In 2008 she was awarded first prize in the BNU Excellent Teaching Achievement. In 2007 she placed first in the Host Teacher of National Education Network Fine Course competition. In 2006 she was the recipient of the second BNU Teaching Masters Award. In 1999 Li Chen was awarded first prize in the Distance Education Research Paper Competition.

**Mingli Fan** (1981–) is a lecturer working in the College of Education, Hebei University, who earned her PhD in the Faculty of Education at Beijing Normal University (BNU) in 2014, where she studied the administrative system of early childhood education in China. Hebei Province was a typical study case in her research work. She has worked in Hebei University for more than ten years, and has written or co-written 5 textbooks, mainly on the play of children, the layout of kindergarten environments, and the administrative management of kindergartens. She has also published 3 books as co-author and 16 journal articles on education in Mainland China.

**Qinhua Zheng** (1977–) is an associate professor of the School of Educational Technology, Beijing Normal University, China. His research interests and expertise include management and quality assurance of distance education, lifelong learning, and learning analytics. He has been involved in many international and national research projects, such as "The Research on Adult Lifelong Learning Competencies in Beijing" Project, supported by the Beijing Education Research Fund, and "Research by large scale statistics and analysis on situation of pilot network colleges of Chinese modern distance education" project, funded by the Chinese Ministry of Education. He has published over 20 pieces of research in these fields. His email: zhengqinhua@bnu.edu.cn.

**Xiaojie Zeng** is an editor of Comparative Education Review published by the Institute of International and Comparative Education at Beijing Normal University and a member of the Educational Theories Publication Commission. Her main research interests are comparative education and school systems. She participated in several national research projects and is the principal investigator of several provincial research projects. She has published over 30 journal articles, such as "The Two Types of School Choice in the Chinese Education System," and has co-authored the book *Family Ecology and Education.*

**Yijin Zhang** is the director of Full Media Center Public Opinion and Review Department of China Education Press, Head of the think tank of China Education Press, editor in chief of the Dandelion Review. His research and practice is in the field of education planning and think tank service, education public opinion analysis and crisis response, teacher growth planning and training guidance, regional education and school development planning, school culture combing and upgrading, reading research and promotion, etc.

**Yimin Gao** is a professor, the vice director of the Institute of International and Comparative Education at Beijing Normal University. His main research interests are comparative education and adult education. He is the general secretary of China Comparative Education Society. There are more than 10 published books, lots of journal articles and many research reports for Ministry of Education.

**Yu Hong** (1984–) was enrolled by Beijing Normal University as an undergraduate in 2003 and achieve his BEd, MEd and Ph.D. degrees while studying and conducting research at BNU for 10 years. His research interests mainly focus on higher education policy and education economics, and he has participated in some country-level research projects such as *"Research of Important Theoretical and Practical Issues in China's Education Resource Allocation (NNSFC, 2013)."* As an assistant research fellow of higher education research institute in Central University of Finance and Economics, he is involved in the elaboration of the CUFE Charter and has accumulated some experience in institutional research.

**Zhiying Nian** (1978–) is the education advisor to the President of Net-Dragon Computer Network Information Technology Co Ltd in Fuzhou, Fujian Province. She is also the associate director of Smart Learning Institute and a researcher from Beijing Institute for the Learning Society, Beijing Normal University.

Her research field covers lifelong learning, comparative higher education and entrepreneurship education. She is working on a number of research projects on lifelong learning collaborated with Beijing Municipan Government, and some international and national research projects as well. She has been involved in two edited books and three translated books published in China, including The Introduction to ProblemBased-Learning from the International Perspective: Theory and Practice by Higher Education Press. She has published over 10 articles in journals. Her email: zhiynian_bnu@126.com.